Nanophotonics and Micro/Nano Optics International Conference (NANOP 2016)

EPJ Web of Conferences Volume 139 (2017)

Paris, France
7 - 9 December 2016

Editors:

Muhammad Faryad

ISBN: 978-1-5108-3834-5

Printed from e-media with permission by:

Curran Associates, Inc.
57 Morehouse Lane
Red Hook, NY 12571

Some format issues inherent in the e-media version may also appear in this print version.

Printed by Curran Associates, Inc. (2017)

For additional information, please contact EDP Sciences – Web of Conferences
at the address below.

EDP Sciences – Web of Conferences
17, Avenue du Hoggar
Parc d'Activité de Courtabœuf
BP 112
F-91944 Les Ulis Cedex A
France

Phone: +33 (0) 1 69 18 75 75
Fax: +33 (0) 1 69 28 84 91

contact-edps@webofconferences.org

Additional copies of this publication are available from:

Curran Associates, Inc.
57 Morehouse Lane
Red Hook, NY 12571 USA
Phone: 845-758-0400
Fax: 845-758-2633
Email: curran@proceedings.com
Web: www.proceedings.com

TABLE OF CONTENTS

EPJ Web of Conferences **139**, 00001 (2017) DOI: 10.1051/epjconf/201713900001
NANOP 2016

Plasmonic Nanostructured Cellular Automata

Emad Alkhazraji[1,*], *A. Ghalib*[2], *K. Manzoor*[2] *and M. A. Alsunaidi*[2]

[1]Jubail Industrial College, Jubail 31961, Saudi Arabia
[2]King Fahd University of Petroleum & Minerals, Dhahran 31261, Saudi Arabia

Abstract. In this work, we have investigated the scattering plasmonic resonance characteristics of silver nanospheres with a geometrical distribution that is modelled by Cellular Automata using time-domain numerical analysis. Cellular Automata are discrete mathematical structures that model different natural phenomena. Two binary one-dimensional Cellular Automata rules are considered to model the nanostructure, namely rule 30 and rule 33. The analysis produces three-dimensional scattering profiles of the entire plasmonic nanostructure. For the Cellular Automaton rule 33, the introduction of more Cellular Automata generations resulted only in slight red and blue shifts in the plasmonic modes with respect to the first generation. On the other hand, while rule 30 introduced significant red shifts in the resonance peaks at early generations, at later generations however, a peculiar effect is witnessed in the scattering profile as new peaks emerge as a feature of the overall Cellular Automata structure rather than the sum of the smaller parts that compose it. We strongly believe that these features that emerge as a result adopting the different 256 Cellular Automata rules as configuration models of nanostructures in different applications and systems might possess a great potential in enhancing their capability, sensitivity, efficiency, and power utilization.

1 Introduction

Cellular Automaton (CA) theory was established and developed in the second half of the last century [1]. A one-dimensional Cellular Automaton is a discrete mathematical structure that is consisted of a sequence of k number of consecutive cells where each cell can take a value between 0 and $k-1$. The initial state of the structure is dubbed as the first generation or generation 0. Thereafter, the value of each cell is updated in accordance with the cell's present value and its neighbouring cells' values as determined by a specific rule set that is depicted by each of the available 256 CA rules as generations progress. The decimal value of each rule is represented in hexadecimal form and then the binary equivalent of this hexadecimal number associates with different combinations of input values of the cells [1, 3]. Cellular Automata had been extensively used in several applications, such as image processing, traffic modelling, and music composition [2]. Furthermore, different CA can model several universal dynamic phenomena and occurrences ranging from DNA sequences to galactic formations [3].

We believe that employing these nature-complying rules as configuration models in different artificial systems and structures, such as sensing, imaging, and energy harvesting applications, can introduce positive enhancements to their capabilities, efficiencies, and energy utilization. In this work, we investigate the scattering plasmonic resonance profiles of silver spherical nano-particles when CA rules are used as models of their geometrical distribution. Thereafter, we characterize the scattering profiles of these CA configurations while observing how they evolve after each generation.

2 Analysis model

The analysis model of this study is based on the numerical time-domain solution of Maxwell's equations for electrodynamics, where scattering light waves inside the nanostructure is simulated and analyzed. As such, a three-dimensional TMz domain was implemented where the electric and magnetic fields are represented in a finite-difference time-domain scheme according to Yee's 3D staggered mesh. To eliminate the effect of reflections from the finite numerical domain, the generalized material-independent perfect matching layer (GMIPML), introduced in [4], was implemented at all boundaries with a thickness of eight mesh cells. In addition, two different techniques have been used in order to isolate and find the scattered fields and to simulate the dispersive nature of the silver nanospheres.

2.1 Total-fields/scattered-fields

Total-Field/Scattered-Field (TF/SF) technique is used to identify the scattered waves of different scattering objects inside the established domain and to isolate them from the source wave fields. Inside the whole domain, a closed region is defined inside which, the targeted object is cantered. Inside this region, both incident source fields and scattered fields, off the target, are present hence, it is called the Total Region. Outside that region, however, by eliminating the incident fields' components, the scattered fields are isolated and tracked as time progresses. This region is called the Scattered Region.

In order to characterize the scattering plasmonic profile of any structure, the poynting vector of the scattered waves are calculated as a function of time at all sides of the three-dimensional domain. Finally, taking the Fourier transform of the scattered power generates the scattering resonance profile of the target where the peaks of the profile represent the scattering plasmon modes of the scattering target.

2.2 Simulation of Dispersive Nature of Silver

The General Algorithm proposed in [5] is used to model the dispersive nature of the plasmonic nano-structures. Following Lorentz–Drude model, the polarization vector P for each pole can be obtained using

$$P^n = \frac{4}{2+\tau_i\Delta t}P^{n-1} - \frac{2-\tau_i\Delta t}{2+\tau_i\Delta t}P^{n-2} + \frac{2\varepsilon_o f_i\omega_i^2\Delta t^2}{2+\tau_i\,\Delta t}E^{n-1} \quad (1)$$

where n represents the time step while τ_i, ω_i, and f_i are the pole parameters. Thereafter, the electric fields are updated using

$$E^n = \frac{D^n - \Sigma_i^N P_i^n}{\varepsilon_0\varepsilon_\infty} \quad (2)$$

where D is the electric flux density and N is the number of poles which is 6 in the case of silver whose corresponding poles' parameters were obtained from [6].

3 Implementation of CA rules

Two binary one-dimensional CA, namely CA 30 and CA 33, have been adopted to model the spatial distribution of the silver nanospheres which have been distributed over a plane that is normal to the direction of propagation of the incident light wave. Each generation is 11 cells-wide where each cell can take either a value of 0 or 1 dictating the absence or the presence of a nanosphere in that cell location, respectively. The radius of the nanospheres was 5 nm with a separation of 1 nm.

3.1 Cellular Automaton 33

Cellular Automaton rule 33 is a simple one-dimensional non-divergent CA rule [7]. Fig.1 shows the first 10 generations of that rule for a width of 11 cells/generation. To follow this CA rule as a model, initially, the first generation was implemented by deploying a single sphere (see inset of Fig. 2) at the top middle of the total region inside the domain. The scattering profile of this configuration, as shown in Fig. 2, is consisted of two resonance wavelengths, which matches that of a typical silver nanosphere [8,9], as represented by the peaks at 257 nm and 378 nm, respectively.

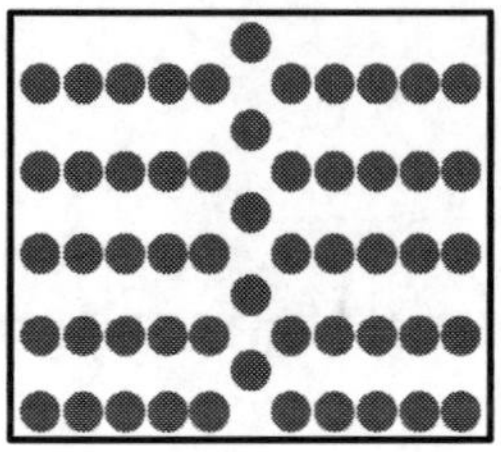

Fig. 1. CA rule 33 after 10 generations of a width of 11 cells.

Thereafter, going in generation-by-generation for 10 generations, more silver nanospheres have been added according to CA rule 33 as shown in Fig. 1. After each generation, the scattering profile had been found to be similar to that of first generation albeit with a slight red or blue shift in both peaks as generations progressed.

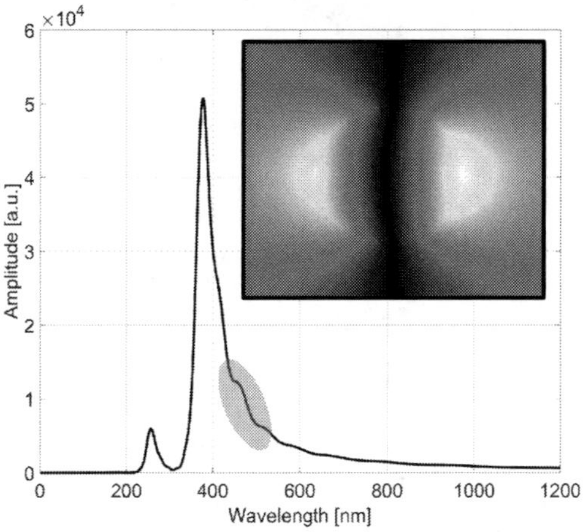

Fig. 2. Scattering profile of the first generation that is consisted of a single nanosphere (inset).

3.2 Cellular Automaton 30

Cellular Automaton 30, much like CA 33, is a one-dimensional CA whose cells can occupy one of two states. The inset in Fig. 3 shows the first eight generations of this rule for a width of 15 cells/generation. More interestingly, this CA rule is mostly known for its chaotic nature as it is classified as a class III CA rule and, hence, is used in random number generators. Moreover, some natural phenomena can be modelled by this rule such as patterns on seashells and other creatures [10].

Starting with an identical first generation to that of CA 33, eight generations of CA 30 were implemented one after another each with a width of 15 cells. Fig. 3 shows a side and a top view (left to right) of the configuration after the eighth generation in plasmonic resonance.

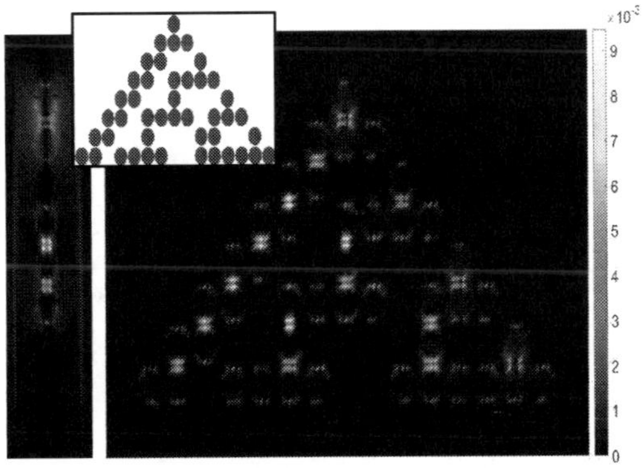

Fig. 3. A side and a top view (left to right) of CA 30 after 8 generations. The inset illustrates the first eight 15-cell-wide generations of that rule.

After adding the second and third generations, the effect on the scattering profile was similar to that of CA 33 where slight shifts were observed in both modes of the profile. However, after introducing the fourth generation, a very peculiar anomaly is witnessed in the scattering profile as a new peak emerges at a wavelength of 463 nm as shown in Fig. 4 by the solid curve. Furthermore, the newly added particles in the fourth generation alone were isolated and their scattering profile was acquired. This was done in order to investigate the newly emerged mode and identify whether its appearance is attributed to the structure itself or if it is merely a sum of its parts. Surprisingly enough, as Fig. 4 demonstrates, the profile of the isolated nanospheres that were introduced in the fourth generation (represented by the dashed curve) matched that of the of the structure after 3 generations (represented by the dotted curve). This indicates that indeed the Cellular Automaton structure itself is introducing its own new features rather than a mere superposition of its composing elements.

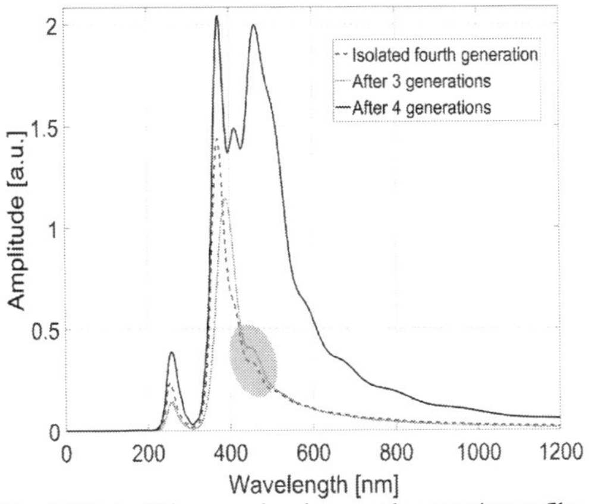

Fig. 4. Side-by-Side comparison between the scattering profiles of the CA 30 structure after 3 generations (dotted curve) and after 4 generations (solid curve), and of the isolated newly added nanoshperes in the fourth generation alone (dashed curve).

Furthermore, in all prior scattering profiles, a small hump was observed (shaded regions in Figs. 2 and 4) around the particular wavelength where the new peak has emerged. We postulate that the Cellular Automaton configuration is enhancing the plasmons of the collective nanospheres at the hump and is enabling it to become a fully-fledged resonance mode.

4 Conclusion

Two Cellular Automata have been implemented as models of distribution of silver nanospheres whose scattering profiles were acquired after each generation. While Cellular Automaton rule 33 is a simple and a systematic rule, which provides simplicity and ease of implementation, it showed a scattering profile matching that of a single silver nanosphere with two resonance modes that blue and red shift as more generations are introduced. On the other hand, Cellular Automaton rule 30 is a chaotic and a divergent rule which introduces difficulties and limitations in design and implementation. Whilst the first few generations showed a similar profile to CA 33, the later generations, however, introduced a new mode that was accredited to the CA 30 configuration itself as it enhanced the plasmonic scattering at that particular mode. A new scattering resonance mode can be directly correlated to providing an extra degree of sensing capabilities in bio-sensing or chemical sensing as some features of the sensing target can be translated by either a shift or change in the response waist size in any of these modes.

With all said and done, there is a total of 254 Cellular Automata besides the two which were used in this work. This study only scratches the surface of the great potential that we believe can be achieved by the adoption of these universal dynamic structures as models in our different artificial applications and systems of nanostructures that can enhance their capabilities, sensitivities, and efficiencies.

References

1. S. Wolfram. "*Cellular Automata as models of complexity*," Nature, **311**, 419-424 (Oct. 1984)
2. D. Das. "*A survey on Cellular Automata and its application,*" Global Trends in Computing and Communication Systems, Berlin Heidelberg: Springer, 753-762 (2012)
3. K. Fujibayashi, R. Hariadi, S. Park, E. Winfree, and S. Murata. "*Toward reliable algorithmic self-assembly of DNA tiles: a fixed-width Cellular Automaton pattern,*" Nano Letters **8(7)**, 1791-1797 (2007)
4. A.P. Zhao, J. Juntunen, A.V. Raisanen. "*Generalized material-independent PML absorbers for the FDTD simulation of electromagnetic waves in arbitrary anisotropic dielectric and magnetic media,*" Microwave and Guided Wave Letters, IEEE, **8(2)**, 52-54 (1998)

5. A. Al-Jabr, M. A. Alsunaidi. "*An efficient time-domain algorithm for the simulation of heterogeneous dispersive structures,*" Journal of Infrared, Millimeter, and Terahertz Waves, **30**, 1226-1233 (2009)
6. A. D. Rakic, A. B. Djurisic, J. M. Elazar, M. L. Majewski. "*Optical properties of metallic films for vertical cavity optoelectronic devices,*" Appl. Opt., **37**, 5271–5283 (1998)
7. W. Eric. "Elementary Cellular Automaton," (2015) URL: www.mathworld.wolfram.com/ElementaryCellularAutomaton.html
8. V. Yannopapas, N.V. Vitanov. "Ultra-subwavelength focusing of light by a monolayer of metallic nanoshells with an adsorbed defect," RRL, **2(6)**, 287-289 (2008)
9. M. K. Schmidt, R. Esteban, J. Sáenz, I. Suárez-Lacalle, S. Mackowski, J. Aizpurua. "Dielectric antennas-a suitable platform for controlling magnetic dipolar emission," Optics express, **20(13)**, 13636-13650 (2012)
10. Weisstein, W. Eric. "Rule 30." (2011) URL: www.mathworld.wolfram.com/Rule30.html

EPJ Web of Conferences **139**, 00002 (2017) DOI: 10.1051/epjconf/201713900002
NANOP 2016

Synthesis, Characterization and Optical Constants of Silicon Oxycarbide

Faisal Ahmed Memon[1,2,*], *Francesco Morichetti*[1], *Muhammad Ishaque Abro*[3], *Giosue Iseni*[4], *Claudio Somaschini*[4], *Umair Aftab*[3], *Andrea Melloni*[1]

[1]Dipartimento di Elettronica, Informazione e Bioingegneria (DEIB), Politecnico di Milano, via Ponzio 34/5, 20133 Milano Italy
[2]Department of Telecommunications Engineering, Mehran University of Engineering & Technology, Jamshoro 76062, Pakistan
[3]Department of Metallurgy and Materials Engineering, Mehran University of Engineering & Technology, Jamshoro 76062, Pakistan
[4]Polifab, Politecnico di Milano, via Giuseppe Colombo 81, 20133 Milano Italy

Abstract. High refractive index glasses are preferred in integrated photonics applications to realize higher integration scale of passive devices. With a refractive index that can be tuned between SiO_2 (1.45) and a-SiC (3.2), silicon oxycarbide SiOC offers this flexibility. In the present work, silicon oxycarbide thin films from 0.1 – 2.0 μm thickness are synthesized by reactive radio frequency magnetron sputtering a silicon carbide SiC target in a controlled argon and oxygen environment. The refractive index n and material extinction coefficient k of the silicon oxycarbide films are acquired with variable angle spectroscopic ellipsometry over the UV-Vis-NIR wavelength range. Keeping argon and oxygen gases in the constant ratio, the refractive index n is found in the range from 1.41 to 1.93 at 600 nm which is almost linearly dependent on RF power of sputtering. The material extinction coefficient k has been estimated to be less than 10^{-4} for the deposited silicon oxycarbide films in the visible and near-infrared wavelength regions. Morphological and structural characterizations with SEM and XRD confirms the amorphous phase of the SiOC films.

1 Introduction

High performance optical waveguide materials must combine low loss with the flexibility to fabricate a variety of passive devices on a single material platform. With a refractive index that can be tuned between silica glass SiO_2 (1.45) and amorphous silicon carbide a-SiC (3.2), silicon oxycarbide SiOC offers this flexibility. However, the methods which were used to produce silicon oxycarbide such as chemical vapor deposition (CVD) [1, 2, 3, 4] and sol-gel pyrolysis [5, 6, 7, 8] introduce hydrogen and high material absorption. In the recent years, the authors in [9,10] showed the deposition of SiOC films with RF sputtering technique and discussed physical and structural characteristics of SiOC thin films. In another study by T. Nakai et. al. [11], low refractive-index SiOC films were developed with rf sputtering and exploited for HD DVD rewritable media. However optical properties (n and k) of the SiOC films over a wide wavelength spectrum ranging from UV to near-infrared are not reported. In this work, we demonstrate fabrication of silicon oxycarbide thin films on Si (100) substrate by reactive RF magnetron sputtering of a SiC target in a controlled argon and oxygen gases environment. The SiOC thin films are characterized by variable angle spectroscopic ellipsometry over the broad UV-VIS-NIR wavelength range to obtain optical constants (refractive index and absorption spectra). The motivation to use sputter deposition comes from the fact that it has potential to prepare high quality material with essentially no hydrogen content at room temperature compared to typical deposition techniques such as CVD [3] which introduce hydrogen and high material absorption coefficient α on the order of 10^5 cm^{-1}. The refractive index n of the deposited SiOC films ranges from 1.41 to 1.93 at 600 nm as a function of RF power level. The extinction coefficient k of all the deposited SiOC films is estimated to be less than 10^{-4} above $\lambda = 1000$ nm. To the best of our knowledge, it is the first time to report optical properties mainly the refractive index n and extinction coefficient k spectra in a broad wavelength range extending from UV (300 nm) to Near IR (1200 nm). Profilometry and scanning electron microscopy (SEM) were used to determine thickness and analyse morphology of the deposited SiOC films, respectively. The phase of the films was examined with X-ray diffraction (XRD).

2 Deposition of SiOC Films

Depositions of silicon oxycarbide thin films were carried out by reactive RF magnetron sputtering of a 99.9% pure 3 inches diameter silicon carbide SiC target which was manufactured by Ajaint International. The sputtering chamber was pumped down to better than 10^{-4} Pa before the argon and oxygen gas flux. High purity argon Ar and oxygen O_2 were used as the plasma and reactive gases in

* Corresponding author: faisalahmed.memon@polimi.it

the sputtering chamber during preparation of amorphous silicon oxycarbide films. The depositions were performed on boron doped p-type Si (100) substrate at room temperature. Prior to deposition, substrates were first cleaned in acetone and then isopropanol in ultrasonic cleaner to remove the contaminants and organic matter on Si substrate surface. In order to clean the SiC target surface, fifteen minutes pre-sputtering without reactive gas was done each time before the actual film deposition. To check repeatability and achieve precise control of the sputter process a large number of SiOC samples were produced by setting the recipe: RF power level from 50 to 450 watts, argon to oxygen gas ratio of 47 sccm and deposition time of 15 minutes. Here we report the results of 6 SiOC films which were deposited as function of increasing RF power level.

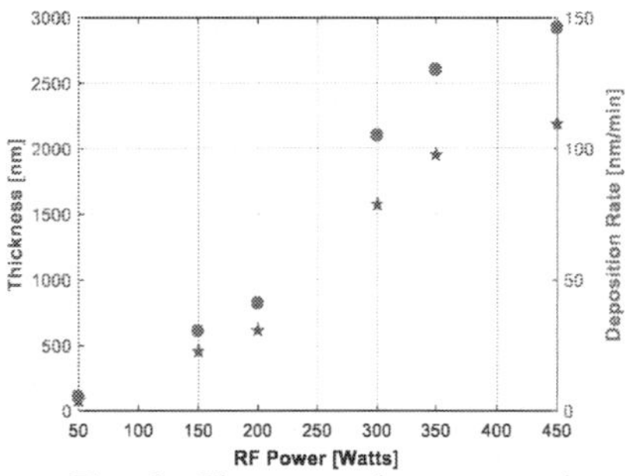

Fig. 1. Blue stars show measured thickness and red circles show deposition rate of the deposited SiOC films as a function of RF power.

The thickness of the SiOC films was measured with KLA Tencor P-15 Surface Profilometer. To determine the thickness with profilometer, kapton tape was used on cleaned substrates as a mask before deposition. The deposited samples were cleaned in acetone by means of ultrasonic device to eliminate the mask. Due to the good quality adhesion of sputtered SiOC films on Si (100) substrates, no wear off or pin holes in SiOC films were observed after ultrasonication. Figure 1 shows the thickness and deposition rate of silicon oxycarbide thin films deposited as a function of RF power from 50 to 450 W. The thickness of the SiOC films was determined to be between 100 nm and 2000 nm. The deposition rate increases with RF power that is varied from 50 to 450 W and a large rate of 150 nm/min has been achieved, which suggests that micron size films can be prepared in a fraction of an hour.

3 Characterization of SiOC Films

The refractive index *n* and extinction coefficient *k* dispersion spectra of the SiOC films were measured with variable angle spectroscopic ellipsometer (VASE J.A. Woollam Inc. USA). The measurement angles used were 72° and 74° which are commonly adopted in [12]. All the SiOC samples were characterized over the wide wavelength range from 300 – 1200 nm which covered UV – Visible – Near IR regions with spectral sampling rate of 10 nm and ellipsometric ψ and Δ data were acquired. A four-layer film/substrate stack comprising surface-roughness, SiOC thin film, native SiO_2 layer, Si substrate as shown in figure (2) was used as a model for our samples and to extract the optical parameters *n* and *k* of the SiOC films. To best fit the data and extract the optical parameters over the entire wavelength range, Tauc-Lorentz (TL) model [13] that is intrinsically Kramers-Kronig [13,14] consistent was used to represent the SiOC film and the obtained extinction coefficient and refractive index values did not require further processing. Jellison and Modine [13] developed the Tauc-Lorentz model which is formulated as,

$$\varepsilon_{n\_T-L} = \varepsilon_{n1} + i\varepsilon_{n2} \quad where \tag{1}$$

$$\varepsilon_{n2} = \begin{cases} \frac{1}{E} \cdot \frac{A_n\, Eo_n C_n\, (E - Eg_n)^2}{(E^2 - Eo_n^2)^2 + C^2E^2} & for\; E > Eg_n \\ 0 & for\; E \le Eg_n \end{cases} \tag{2}$$

and

$$\varepsilon_{n1} = \frac{2}{\pi} P \int_{Eg_n}^{\infty} \frac{\xi .\, \varepsilon_{n2}(\xi)}{\xi^2 - E^2}\, d\xi \tag{3}$$

Equation (1) is the combination of real (2) and imaginary parts (3) of the dielectric function of Tauc-Lorentz oscillator. In equation (2), A_n is the strength of the $\varepsilon_{n2}(E)$ peak in eV, C_n is the broadening of the peak in eV, Eo_n is the central energy of the peak in eV and Eg_n is the bandgap energy in eV. Since Eg_n is declared as a fit parameter in the TL oscillator to model the ellipsometric ψ and Δ of the SiOC films, the value does not represent the actual bandgap of the material. The real part ε_{n1} (3) of the dielectric function (1) is derived from the expression of ε_{n2} (2) using the Krameres-Kronig integration [13,14,15]. In equation (3), P is the Cauchy principal value containing the residues of the integral at poles located on lower half of the complex plane and along the real axis. WVASE® [15] uses complete analytical solution to the Kramers-Kronig integral as given in [15].

High resolution images of silicon oxycarbide films cross section were obtained with a LEO 1525 scanning electron microscope (SEM). To avoid electrons charging effect, a thin layer of platinum (≈ 2 nm) was deposited on SiOC films. The electrons were accelerated at a voltage of 10 kV and the signal was collected from in-lens detector keeping the distance around 6 mm between the SiOC sample and the electron gun. The microstructure of the SiOC films was examined from x-ray diffraction patterns measured over Range 2θ 20 – 50 degrees with XRD D8 Advance (Bruker Corp).

4 Results and Discussion

The optical properties of the silicon oxycarbide films grown on Si subtstrate as a function of RF power have been characterized by spectroscopic ellipsometry. The ellipsometric ψ and Δ data acquired over the wide wavelength range from 300 to 1200 nm for each SiOC film have been modelled with a four-layer model comprising surface roughness/Tauc-Lorentz/native-SiO_2/Si layers as illustrated in figure 2. Since ellipsometry is very sensitive to native silica and surface roughness conditions, the 2 nm SiO_2 and surface roughness layers were added to improve fit and MSE. The optical constants of Si substrate and SiO_2 were used from the WVASE® database [15]. The surface roughness SR layer based on Bruggemann and Mawxell-Garnett effective medium approximation (EMA) [16] composed of 50 % material and 50 % voids is used to simulate the surface roughness of SiOC films. The SR layer is ellipsometry equivalent of AFM and gives peak-to-valley roughness, not the rms roughness. The parameters of the Tauc-Lorentz oscillator described earlier were declared fit in the model shown in figure 2(a) and regression analysis algorithm was run to minimize the global mean squared error (MSE). Figure 2(b) shows an example of fitting of ellipsometric ψ and Δ spectra of the SiOC film deposited at 300 W over the broad wavelength range 300 - 1200 nm at one measurement angle 74° for the sake of brevity.

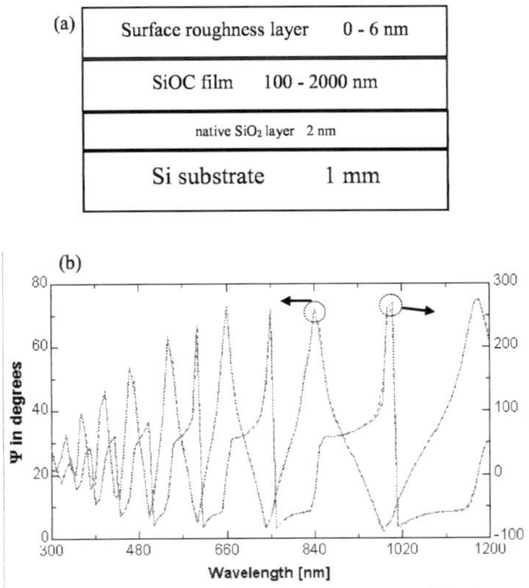

Fig. 2. (a) four-layer stack to model the ellipsometric ψ and Δ data (b) ellipsometric ψ and Δ data (green and blue dotted lines) of SiOC film deposited at RF power 300 W acquired at measurement angle of 72° and model fit (red solid lines)

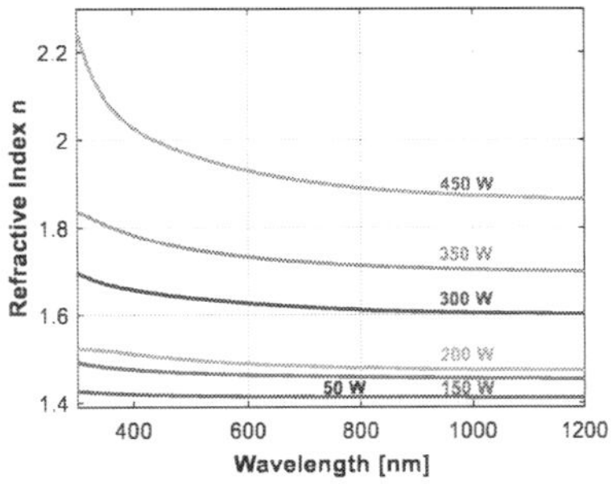

Fig. 3. Refractive index *n* spectra of SiOC films deposited as a function of RF power 50 - 450 W

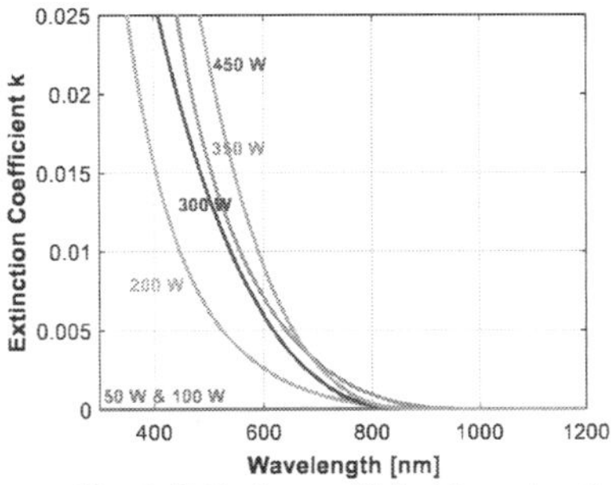

Fig. 4. Extinction coefficient *k* spectra of SiOC films deposited as a function of RF power 50 - 450 W

Figures 3 and 4 show the refractive index *n* and material absorption (extinction coefficient) *k* spectra of the SiOC films deposited by increasing RF power level from 50 to 450 W. As shown in figure 3, the refractive index spectral curves are decreasing as a function of wavelength from UV to Near-IR region which demonstrate a typical dielectric medium behaviour. The refractive index *n* of the deposited SiOC films ranges from 1.41 to 1.93 at visible light wavelength λ = 600 nm. The extinction coefficient *k* in all the SiOC films is less than 10^{-4} above λ = 1000 nm. In the visible region, the SiOC films show better transparency as compared to SiOC films deposited with CVD [3]. The roughness of the SiOC films deposited at RF power level of 450 W is found to be around 6 nm by fitting SR layer and it reduces to almost nothing with decreasing RF power level to 50 W.

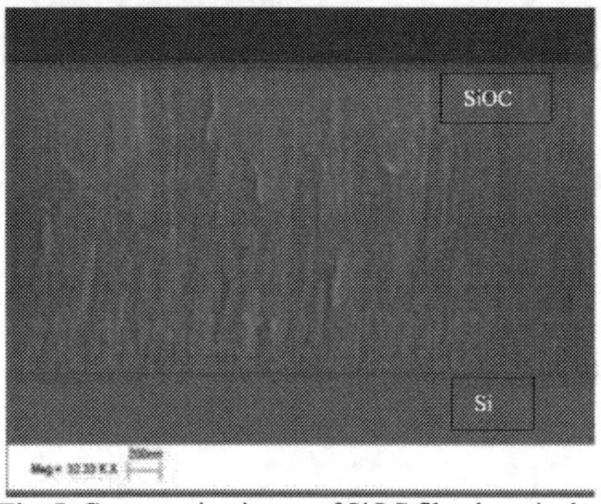

Fig. 5. Cross section image of SiOC film deposited at 350 W obtained with SEM

The cross section of the SiOC films was obtained with scanning electron microscope SEM. A high resolution image of the SiOC film deposited at RF power level of 350 W is shown in figure 5. The SiOC film seems to possess columnar structure at higher RF power levels.

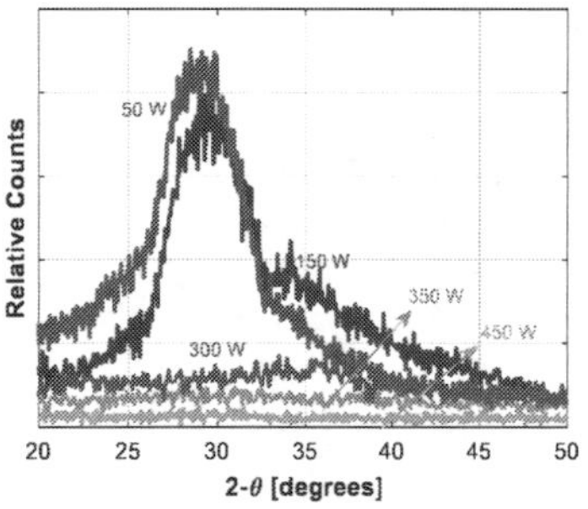

Fig. 6. XRD patterns of SiOC films deposited as a function of RF power (50 – 450 W)

X-ray diffraction patterns of the SiOC films deposited as function of RF power (50 – 450 W) are shown in figure 6. Since no sharp peaks have been detected in the XRD spectra, the SiOC films seems to be amorphous in phase. The two broad spectral peaks visible at lower RF power levels (50 and 150 W) in figure 6 are due to the presence of amorphous silicon dioxide in the SiOC films deposited at lower RF power levels, as they have greater oxygen percentage. The results are in agreement with literature [9].

5 Conclusion

In this work, silicon oxycarbide films have been deposited by reactive RF magnetron sputtering technique on Si substrate. The optical properties of the SiOC films deposited as a function of RF power are comprehensively studied with variable angle spectroscopic ellipsometry over the broad UV-Vis-Near IR wavelength range 300 – 1200 nm. The ellipsometric ψ and Δ data of the SiOC films have been fitted with lower MSE which demonstrates the accuracy of obtained optical constants. The Kramers-Kronig consistent Tauc-Lorentz model has been used to describe SiOC films and extract its optical parameters. The refractive index n of the deposited SiOC films range from 1.41 to 1.93 at 600 nm as a function of RF power. The material extinction coefficient k in all the deposited SiOC films has been estimated lower than 10^{-4} above λ = 1000 nm. In the visible spectrum region, SiOC films also show good transparency. Structural characterization with XRD suggested amorphous phase of the deposited SiOC films. In comparison to CVD [3], SiOC films have been synthesized using sputtering technique with higher deposition rate and lower extinction coefficient k values. Results on SiOC films are promising for integrated photonics applications.

F. A. Memon acknowledges the financial support in his PhD from European Union under Erasmus Mundus 'LEADERS' project.

References

1. Grill, Alfred, J. Appl. Phys. **93**, 1785-1790 (2003)
2. H. J. Kim, Q. Shao, Y. H. Kim, Surf. and Coat. Tech., **171**, 39-45, (2003)
3. S. Gallis, V. Nikas, E. Eisenbraun, M. Huang and A.E. Kaloyeros, J. Appl. Phys. **102,** 024302, (2007)
4. R. J. Martín, R. Gago, V. Torres, P. Fernández, U. Kreissig and J.M. Martínez Duart, Thin Sol. Films, **515**, 2493-2496, (2006)
5. V. S. Pradeep, M. G. Zajac, R. Riedel and G. D. Soraru, Electrochimica Acta, **119**, 78-85, (2014)
6. A. Karakuscu, R. Guider, L. Pavesi, and G. D. Sorarù, J. Amer. Ceram. Soc. **92,** 2969–2974, (2009)
7. Sorarù, G. D., D'Andrea, G., Campostrini, R., Babonneau, F. and Mariotto, G., J. Amer. Ceram. Soc., **78**, 379–387, (1995)
8. G. Das, P. Bettotti, L. Ferraioli, R. Raj, G. Mariotto, L. Pavesi and G. D. Sorarù, Vib. Spectr., **45**, 61-68, (2007)
9. J. V. Ryan and C. G. Pantano, J. Vac. Sc. & Technology A, **25**, 153-159 (2007)
10. P. Du, X. Wang, I.-K. Lin, X. Zhang, Sensors and Actuators A 176, 90–98 (2012)
11. T. Nakai, K. Yusu, Y. Satoh, and S. Ashida. J. Appl. Phys. 45, 1447 (2006)
12. J. I. Larruquert, A. P. Pérez-Marín, S. García-Cortés, L. R. Marcos, J. A. Aznárez, and J. A. Méndez, J. Opt. Soc. Am. A **28**, 2340-2345 (2011)
13. G. E. Jellison, and F. A. Modine., Appl. Phys. Lett. **69**, 371-373 (1996), Erratum, Appl. Phys. Lett. **69**, 2137 (1996)
14. Tauc-Lorentz Dispersion formula, Technical note available at www.horiba.com
15. Guide to using WVASE® spectroscopic ellipsometry data acquisition and analysis

software, J. A. Woollam Co., Inc., chap.4, p-16-20 & chap. 17, p-21-31 (2012)
16. D. A. G. Bruggeman, Ann. Physik, **24**, 636 (1935)

High sensitive photonic crystal multiplexed biosensor array using H0 sandwiched cavities

Safia Arafa*, *Mohamed* Bouchemat, *Touraya* Bouchemat, and *Ahlem* Benmerkhi
Electronics Department, Université des frères Mentouri, Constantine 1, 25017 Algeria

Abstract. We theoretically investigate a high sensitive photonic crystal integrated biosensor array structure which is potentially used for label-free multiplexed sensing. The proposed device consists of an array of three sandwiched H0 cavities patterned above silicon on insulator (SOI) substrate; each cavity has been designed for different cavity spacing and different resonant wavelength. Results obtained by performing finite-difference time-domain (FDTD) simulations, indicate that the response of each detection unit shifts independently in terms of refractive index variations. The optimized design makes possible the combination of sensing as a function of location, as well as a function of time in the same platform. A refractive index sensitivity of 520nm/RIU and a quality factor over 10^4 are both achieved with an accompanied crosstalk of less than -26 dB. In addition, the device presents an improved detection limit (DL) of $1.24.10^{-6}$ RIU and a wide measurement range. These features make the designed device a promising element for performing label-free multiplexed detection in monolithic substrate for medical diagnostics and environmental monitoring.

1 Introduction

Label-free optical biosensors have recently attracted increasing interest for high-performance quantitative measurements without heterogeneity labeling provided by the fluorescent techniques. In particular, photonic crystal (PhC) biosensors, because of their strong capability to manipulate light propagation, their miniaturized size, their high spectral sensitivity and their minimal sample preparation, have generated substantial attention as an attractive candidate for available label-free systems [1–15]. The performance of these biosensors is evaluated by the ability to detect the small change of refractive index (RI). For resonant biosensors, this ability is expressed in terms of quality (Q) factor and sensitivity (S). The Q-factor is defined as $Q=\lambda_0/\Delta\lambda$, where λ_0 is resonance wavelength and $\Delta\lambda$ is the line-width of the resonance peak, It represents the strength of light confined in a cavity. The sensitivity of a device is reflected by the magnitude of light-matter interaction. Among all the different biosensor designs that have been developed, the biosensors based on PC coupled waveguide micro-cavities systems have attracted significant interests due to their compact size, high Q-factor and high S [10-15].

However, the design drawback of these biosensors is that they typically operate as point or single detection unit and the number of targets which can be screened for at once is relatively small. Recently, in order to overcome this drawback and to detect different analytes simultaneously on a single platform, sensor arrays using PhC have been developed. Examples of such systems include that of Mandal *et al.* [16] who have firstly proposed a nanoscale optofluidic sensor array based on 1D PhC silicon waveguide for label-free detection of biomolecular interactions in aqueous environments. Yang *et al.* [17] who have demonstrated a nanoscale PhC sensor array on monolithic substrates using a side-coupled resonant cavity. Olyaee *et al.* [18] have also demonstrated a sensor array based on side coupled H1 cavities to PhC waveguide where sensitivity of 165.45nm/RIU was obtained. In order to efficiently increase the integration density of sensor array and restrain the crosstalk between adjacent PhC sensors, Yang et al. [19] have reported the investigation of an integrated sensor array based on H0 cavities side coupled to parallel output waveguides with crosstalk value lower than -4dB. Further, Liu et al. [20] have proposed an optical PhC sensor based on a channel-drop configuration with a reflector and an additional PhC cavity. Simulation and experimental results showed a good agreement with RI sensitivity of 153nm/RIU. Recently, In order to reduce the crosstalk effect Huang *et al* [21] have proposed a label-free integrated ring-slot sensor array structure that exhibits a Q factor of 11.000 and a crosstalk value of less than -25.8 dB in water environment.

This work aims to investigate a high sensitive PhC integrated biosensor array structure, which is potentially used for label-free multiplexed sensing in aqueous environment. In order to further improve the detection level and simultaneously restrain the crosstalk value between the adjacent biosensors, the well-known technique of wavelength division multiplexing to

* Corresponding author: arafa.safia@gmail.com

spatially separate the integrated biosensor is used. Results obtained by performing FDTD simulations indicate that the response of each sensor unit to the RI variation is completely independent and the resonance spacing between the adjacent sensors is wide enough to ensure high multiplexed detection.

2 PHC sandwiched cavity design

In order to efficiently enhance the integration density of biosensor array and restrain crosstalk value between the adjacent detection units, the sandwiched cavity design is proposed where the operation of the proposed device as a biosensor is based on a cavity coupled waveguide system. The two waveguides are obtained by removing one row of air holes in the x direction. They are used to couple light in and out of the PhC cavity.

As shown in Fig. 1, the proposed structure consists of a hexagonal lattice of air holes (*r*=180.5 nm) patterned periodically with a lattice constant *a* (*a*=475 nm) on a silicon on insulator (SOI) slab with an effective index of n_{eff}=2.87 which corresponds to the effective index of the fundamental guided TE mode in a 230 nm thick silicon slab on silicon dioxide at the wavelength of 1550 nm. Perfectly matched layers (PML) conditions have been considered in the calculations to ensure no back reflection in the limit of the analyzed region [22].

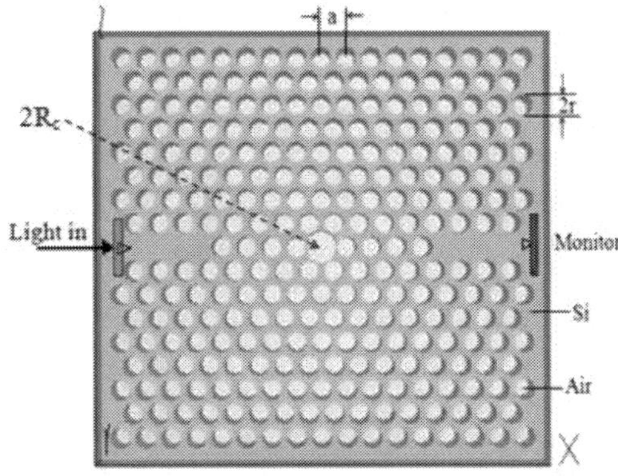

Fig. 1. Schematic diagram of the proposed PhC H0 cavity coupled to an input and output PhC waveguides. The resonator-waveguide system is patterned above a SO I substrate.

In the simulation process, a TE polarized Gaussian optical pulse, covering the whole frequency-range of interest, is launched at the input port to excite the cavity modes. A power monitor was placed at the end of the output waveguide to measure transmitted signal. For improving accuracy in the simulation, the 2D-FDTD analysis from the commercial RSoft software was carried out with a grid size of 0.01.

Obtained high RI sensitivity S (Δλ/Δn), which is defined as the ratio of the shift in the wavelength (Δλ) to the change in the RI due to analyte infiltration (Δn), depends largely on the increase of both the sensing surface and the Q factor value which increases the ability of the device to detect not only small concentrations, but also small changes in the analyte concentration. This can be achieved by increasing the radius of the central cavity hole (R_c).

The variation tendencies of the Q factor and the resonance wavelengths according to Rc change from 0.579*a* to 0.621*a* are plotted in Fig.2. According to the simulation results, we can note that, by increasing the radius of the central defect hole, the resonant peak shift to shorter wavelength (blue-shift) due to the decrease of the high-dielectric material in the cavity region. Similarly the Q factor gets promoted with the increase of R_c followed by a slight decrease and as seen in Fig.2 the air hole radius of Rc = 0.603*a* leads to an optimal design with a Q factor of 92963.

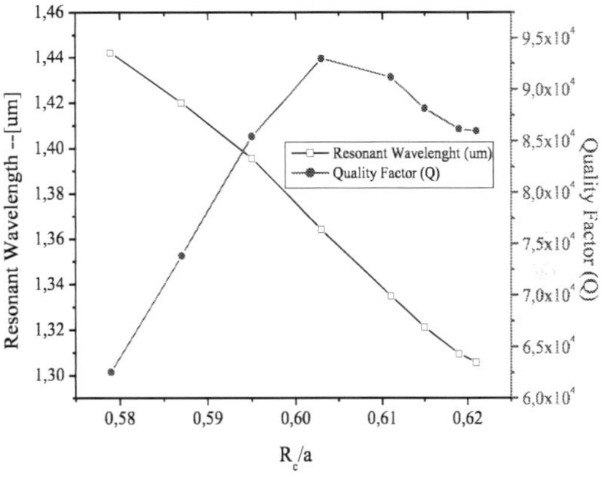

Fig. 2. The variation of the resonant wavelength and Q factor according to the change of the central hole radius.

It is essential to note that for biosensing applications, the magnitude of the resonant wavelength shift is dependent on the combination of many factors such as the number of functionalized holes and the effective refractive index change of targets.

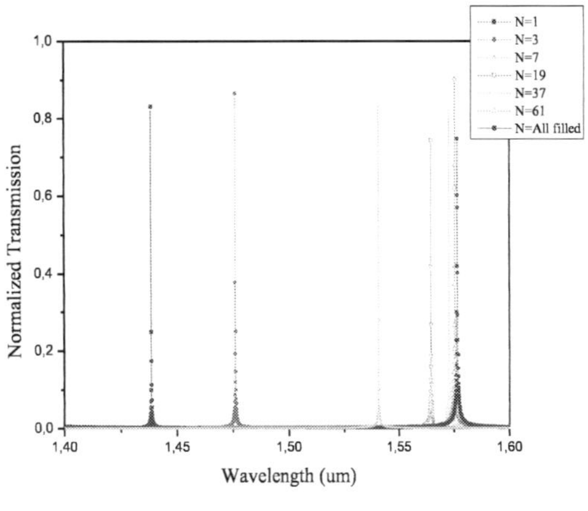

Fig. 3. Transmission spectra with different number of functionalized holes (N).

DOI: 10.1051/epjconf/201713900003

To study the dependence of the sensitivity on the number of functionalized holes, we varied the number of holes being functionalized (N) (filled with the analyte (DI-Water n=1.33) including the central defect hole and the surrounding holes around the resonant cavity.
The transmission spectra corresponding to the different infiltration cases (N varies from 1 hole being functionalized to the total filling of the PhC structure) are depicted in Fig.3. As can be seen in this figure, the resonant wavelength shift gets larger as the number of functionalized holes increases, which is due to the reduced effective refractive index between the infiltrated holes and the semiconductor membrane [23], as a result, the sensitivity becomes higher.

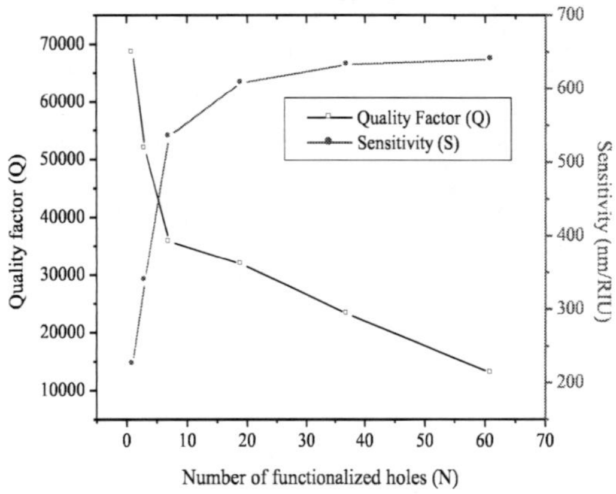

Fig. 4. Quality factor and refractive index sensitivity (S variations as a function of the number of functionalize circular holes (N).

The variation tendencies of the Q factor and sensitivity as a function of the change in the number o functionalized holes (N) that corresponds to the differen local infiltration cases are shown in Fig. 4.

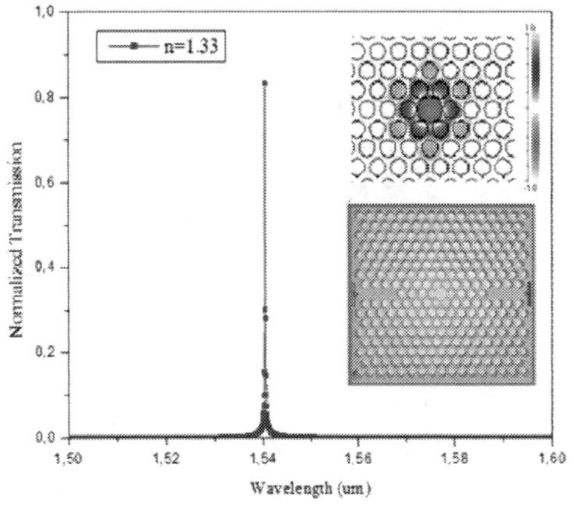

Fig. 5. The transmission spectra of the optimized PhC biosensor in water environment .the electric field distribution in x-y plane and the optimal structure design with the light blue functionalized area are represented in the inset figures.

As indicated in this figure, the sensitivity is enhanced by increasing the number of functionalized holes, whereas the Q factor is declined. Therefore, it has been necessary to choose a trade-off between the Q factor and the RI sensitivity. According to the obtained results, The optimal Q factor of 35718 and the improved sensitivity of 534.54 nm/RIU are both achieved for N=7 where the resonance peak is located at 1504.5 nm (Fig.5). These values present a large figure of merit compared to the finding of previous similar studies [13-15].
Further, we research the detection limit of refractive index, which is an important parameter in sensing applications. The detection limit of refractive index changes (DL) is expressed as follows [24]:

$$DL = \frac{\lambda_0}{10QS} \quad (1)$$

Where λ_0 represents the resonant wavelength, which is equal to 1504.5 nm, S is the RI sensitivity and Q means the quality factor. Based on the above results the calculated detection limit of the proposed biosensor is about $7.88\text{x}10^{-6}$ RIU.

3 PhC biosensors array design

Based on the design and discussion about the use of the sandwiched cavity as a RI biosensor in the above section, we demonstrate a PhC biosensor array with different three cavity spacing's and different resonant wavelengths integrated into a multiplexed array structure.

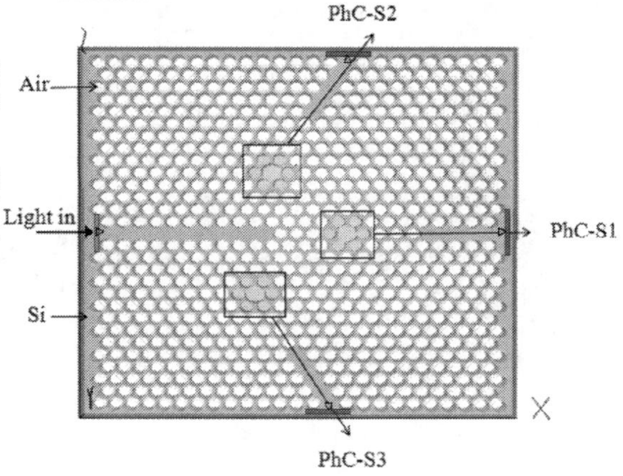

Fig. 6. Schematic diagram of the proposed PhC integrated biosensor array based on H0 sandwiched cavities. The light blue holes indicate the sensing area of each detection unit.

As shown in Fig.6, the specific parameters of the PhC biosensor array are set as follows: PhC-S1: R_{c1}=285; PhC-S2: R_{c2}=291, 5 nm; PhC-S3: R_{c3}=278, 5 nm. As it can be seen in Fig.7, the transmission efficiencies have response peaks at 1517.3 nm, 1547.2 nm and 1575.3 nm in water environment (RI = 1.33) with respect to the green, blue and red solid lines. It can also be noted that the three resonant wavelengths lie in the telecommunication wavelength region with an appropriate resonant spacing and as expected they exhibit high transmission efficiency.

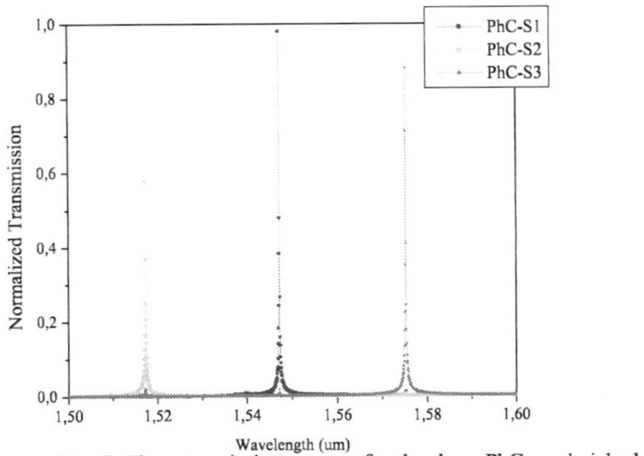

Fig. 7. The transmission spectra for the three PhC sandwiched cavities in water environment (RI=1.33).

The steady-state electric field profile for the fundamental TE-like mode propagation through the three cavities is shown in Fig.8. It can be clearly seen that the light is strongly confined within each resonant cavity, which means that the photon lifetime is enhanced and therefore higher Q factors can be achieved. The calculated Q factors of the three peaks are equal to 44360, 28710 and 25029 respectively. These values showed an improvement in respect to other reports [17-19, 21].

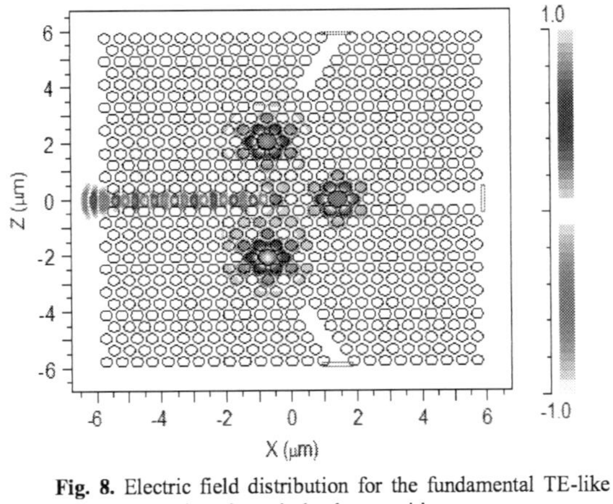

Fig. 8. Electric field distribution for the fundamental TE-like mode propagation, through the three cavities.

3.1. Sensing properties analysis of the integrated PhC biosensors array

To analyze the RI sensitivity of the PhC integrated sensor array in aqueous environment, the RI of the functionalized area of three biosensors is changed, respectively. Fig.9 shows the composed transmission spectra of the three sandwiched cavities (PhC S1, PhC S2, PhC S3) when the functionalized area of one biosensor is under RI changes and the others aren't.

As it can be seen from this figure, the shift occurs only with the functionalized biosensors while the others remain completely unchanged. It can also be noted that the spectral position of the resonating peak detected at the end of output waveguides shifts towards longer wavelengths (red-shift) as the RI of the biosensor functionalized area is increased by 1.330, 1.335, 1.340, 1.345, 1.350, 1.355, 1.360, 1.365, 1.370, 1.377, respectively, which is in good agreement with the previous works [21], [25].

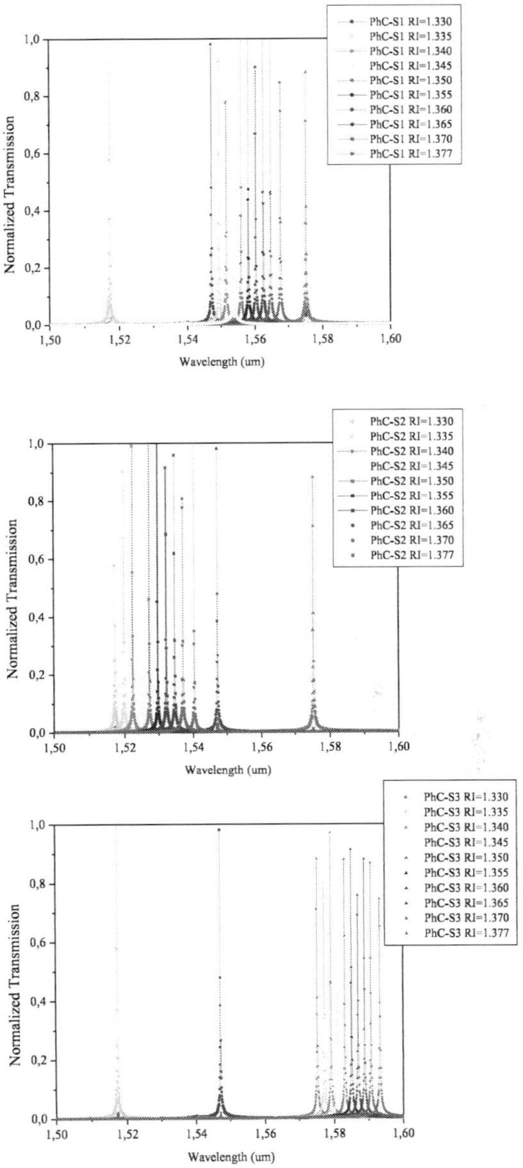

Fig. 9. The transmission spectra of the three sandwiched cavities (PhC S1, PhC S2, PhC S3) when the functionalized area of one sensor is under refractive index variations and the others are not.

Fig. 10 shows the resonance shift of each detection unit as a function of the RI variations. The colored shapes indicate the resonant wavelength shift values (Δλ) as a function of the RI variations (Δn), obtained from FDTD simulations and the red solid line is a linear fit. According to these results, the calculated RI sensitivities (S=Δλ/Δn) of the three biosensors are S1=440 nm/RIU, S2=520 nm/RIU and S3=400 nm/RIU, respectively. Based on Equation (1) the calculated detection limit of the proposed biosensor array is about 6.58x 10-6 RIU. These results are favorably comparable to the similar reported works [19, 21].

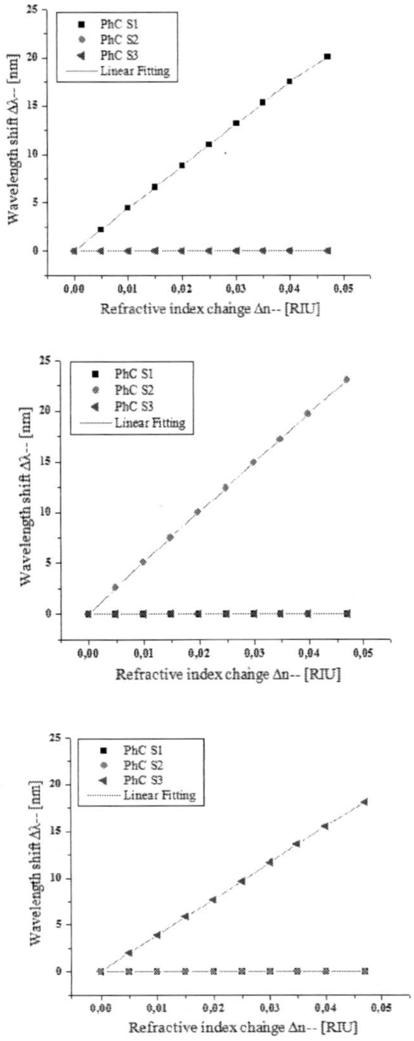

Fig. 10. Red shift in the resonant wavelength Δλ , as a function of the RI increases in the functionalized area and its associated linear fits (solid red line)

3.2 Crosstalk analysis of the integrated PhC biosensors array

The crosstalk between each biosensor unit is an important index for evaluating the component performances.

In this section, the crosstalk between each other adjacent detection unit is analyzed by performing a detailed simulation and by calculating the extinction ratio (ER) which is defined by the following equation: [21]

$$ER = 10\,log\frac{T}{T_i} \quad (2)$$

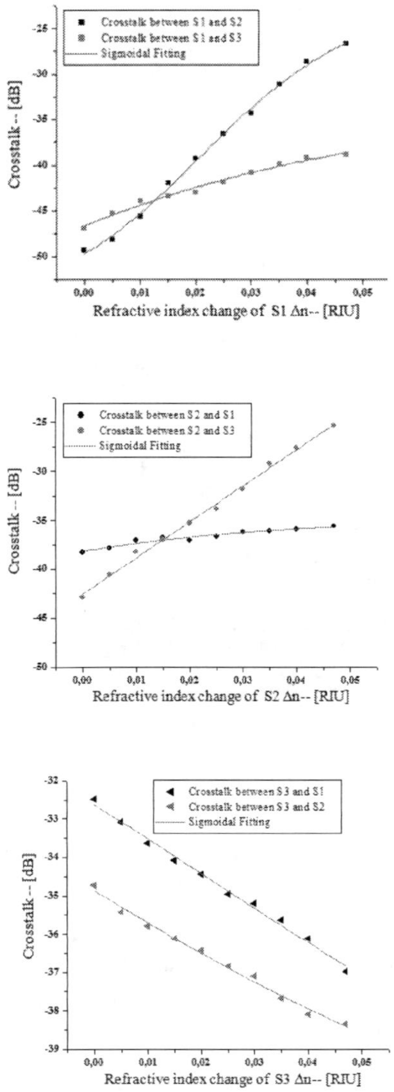

Fig. 11. The crosstalk (dB) between each other adjacent biosensor unit in the proposed biosensor array, when (a) PhC-S1, (b) PhC-S2, and (c) PhC-S3 is under the refractive index variations, respectively and the other biosensors are not.

Where:
T: The maximum transmission efficiency of one sensor at the resonant frequency ω_0 when the RI sensing area of this sensor is varied;
Ti: The minimum transmission efficiency of other adjacent sensors at the same the resonant frequency;
The calculated crosstalk between each other adjacent biosensors (PhC-S1, PhC-S2 and PhC-S3) in the proposed structure is shown in Fig. 10. The crosstalk is calculated when one sensor is exposed to RI variations and the others are not. As seen in this figure the crosstalk between each other adjacent sensor unit in the proposed PhC biosensor array lower than -26dB is observed.
The mean crosstalk value for the three sensors has been estimated to be lower than −26.12 dB. The obtained value has shown a great improvement compared to recent reported works [19, 26]. Furthermore, this value is in a good agreement with the results shown in [21], where the structural design proposed in this study is less complex. This low crosstalk value could potentially be very useful for monolithically integrated and multiplexed sensing technology.

4 Conclusion

In summary, a high sensitive PhC multiplexed biosensor array for label free biodetection has been reported and theoretically demonstrated. The parameters of each biosensor unit have been determined by adjusting the diameter of the central cavity hole, so that, the spectral response of each detection unit appears in telecom range. Simulation results indicated that the response of each biosensor to the RI variations was completely independent and the resonance spacing's between the adjacent ones were wide enough to ensure high parallel detection with an extremely low crosstalk value of -26.3 dB. Considering water absorption at telecom wavelength range, a RI sensitivity of 530 nm/RIU and a Q factor of 44360 have been both achieved with a detection limit of less than 6.57×10^{-6} RIU. The specific results suggest that the proposed device is a promising candidate for performing label-free multiplexed detection in monolithic platform.

References

1. E. Chow, A. Grot, L.W. Mirkarimi, M. Sigalas, G. Girolami, Opt. Lett, **29,** 1093-1095 (2004).
2. Y. Zhang, Y. Zhao, Q. Wang, Sens. Actuators B Chem, **209**, 431–437, (2015).
3. A. Benmerkhi, M. Bouchemat, T. Bouchemat, Optik, **127**, 5682–5687 (2016).
4. A. Benmerkhi, M. Bouchemat, T. Bouchemat, Photon. Nanostr. Fundam. Appl, **20**, 7-17, (2016).
5. N. Skivesen, A. Tetu, M. Kristensen, J. Kjems, L.H. Frandsen, P.I. Borel, Opt. Express, **15**, 3169-3176, (2007).
6. H.S. Dutta, S. Pal, Opt. Quantum Electron, **45**, 907–917, (2013).
7. H.S. Dutta, A.K. Goyal, S. Pal, J.Nanophotonics, **8**, 083088–083088, (2014).
8. F. Bougriou, T. Bouchemat, M. Bouchemat, N. Paraire, Eur. Phys. J. Appl. Phys, **62**, 11201-11206, (2013).
9. Y. Zhang, Y. Zhao, Q. Wang, Sens. Actuators B Chem, **184**, 179–188, (2013).
10. X. Wang, Z. Xu, N. Lu, J.Zhu, G. Jin, Opt. Commun, **281**, 1725–1731, (2008).
11. L. Huang, H. Tian, D. Yang, J. Zhou, Q. Liu, P. Zhang, Y. Ji, Opt. Commun,**332**,42-49,(2014).
12. S. Arafa, M. Bouchemat, T. Bouchemat, A. Benmerkhi, A. Hocini, Opt. Commun, **384**, 93-100, (2017).
13. A. Harhouz, A. Hocini, J. Electromagn. Wave Appl, **29**,659–667, (2015).
14. J. Zhou, H. Tian, D. Yang, Q. Liu, Y. Ji, Opt. Commun, **330**, 175–183, (2014).
15. L. Huang, H. Tian, J. Zhou, Q. Liu, P. Zhang, Y. Ji, Opt. Commun, **335**, 73–77, (2015).
16. S. Mandal, D.Erickson, Opt. Express, **16**, 1623–1631, (2008).
17. D. Yang, H. Tian, Y. Ji, Opt. Express, **19**, 20023–20034, (2011).
18. S. Olyaee, S. Najafgholinezhad, Appl. Optics, **52**, 7206-7213, (2013).
19. D. Yang, H. Tian, Y. Ji, IEEE Photonics J, **6**, 1–7 (2014).
20. Y. Liu, H.W.M. Salemink, J. Lightwave Technol, **33**, 3672–3678, (2015).
21. L. Huang, H. Tian, J. Zhou, Y. Ji, Sensors, **14**, 15658–15668 (2014).
22. J. P. BerengerJ. Comput. Phys.**114**, 185–200 (1994).
23. A. Faraon, D. Englund, I. Fushman, J. Vučković, N. Stolz, P. Petroff Appl. Phys. Lett. **90**, 213110–213113, (2007).
24. I.M. White, X. Fan, Opt. Express, **16**, 1020–1028, (2008).
25. S. Arafa, M.Bouchemat, T. Bouchemat, A. Benmerkhi, Optik, **131**, 49-57, (2017).
26. P. Zhang, H. Tian, D. Yang, Q. Liu, J. Zhou, L. Huang, Y. Ji, Opt. Commun, **355**, 331–336, (2015).

EPJ Web of Conferences **139**, 00004 (2017) DOI: 10.1051/epjconf/201713900004
NANOP 2016

Time resolved spectroscopic investigation of SiD_2 + D_2: kinetic study

Najem A. Al-Rubaiey[1],*, *Robin* Walsh[2]

[1]Petroleum Technology Department, University of Technology, P.O. Box 35109, Baghdad, Iraq
[2]Chemistry Department, University of Reading, Reading, UK

Abstract. Silylenes (silanediyls) have made an important impact on organosilicon chemistry even if it is of more recent foundation than carbenes in organic chemistry and much less complete. These species are highly reactive intermediates. They play a central role in the chemical vapour deposition (CVD) of various silicon-containing thin films which have a technological importance in microelectronics as well as in the dry etching processes of silicon wafers. Spectroscopic methods have been developed to observe these species, a necessary pre-requisite to their direct monitoring. In this work, deuterated phenylsilane precursor, $PhSiD_3$ was chosen for SiD_2 because its analogue phenylsilane, $PhSiH_3$ proved to be a good precursor for SiH_2 and the high quality decay signals observed revealed that SiD_2 be readily detected from $PhSiD_3$ and that if other decomposition pathways (e.g. PhSiD + D_2) are occurring, they do not effect measurements of the rate constants for SiD_2. The absorption spectrum of SiD_2 formed from the flash photolysis of a mixture of $PhSiD_3$ and SF_6 at 193nm were found in the region 17384-17391 cm^{-1} with strong band at 17387.07 cm^{-1}. This single rotational line of $^{P}Q_1$ was chosen to monitor SiD_2 removal. Time-resolved studies of SiD_2 have been carried out to obtain rate constants for its bimolecular reactions with D_2. The reactions were studied over the pressure range 5-100 Torr (in SF_6 bath gas) at four temperatures in the range 298-498K. Single decay from 10 photolysis laser shots were averaged and found to give reasonable first-order kinetics fits. Second order kinetics were obtained by pressure dependence of the pseudo first order decay constants and substance D_2 pressures within experimental error. The reaction was found to be weakly pressure dependent at all temperatures, consistent with a third-body mediated association process. In addition, SiH_2+ H_2 reaction is approximately ca. 60% faster than SiD_2+D_2 reaction. Theoretical extrapolations (using Lindemann-Hinshelwood model and Rice, Ramsperger, Kassel and Marcus (RRKM) theory) were also carried out and obtained data fitted the Arrhenius equations.

1 Introduction

The desire to understand the mechanistic details of the reactions of silylene, SiH_2 prompted us to extend our studies to include isotope effect experiments.[1,2]. The only kinetic information available for deuterated silylene, SiD_2 was that for the reaction with H_2. Mason *et al.*[3] reported a rate constant of $3.8\pm0.2x10^{-12}$ $cm^3molecule^{-1}s^{-1}$ at 5 Torr total pressure in which the reaction found to be pressure independent in the range 2-100 Torr. The possibility of studies with SiD_2 offers the opportunity to give new insights into reaction system previously studied with SiH_2.

For the reaction of SiH_2 with D_2 at room temperature,Jasinski reported a rate constant of $2.6x10^{-12}$ $cm^3molecule^{-1}s^{-1}$[4]. This was the first reported measurement for this reaction. It compares with a value of $3.2x10^{-12}$ $cm^3molecule^{-1}s^{-1}$ for the rate constant for the reaction of SiH_2+H_2 obtained by extrapolation to infinite pressure of a pressure dependent process[5,6]. These high values suggest only a small isotope effect and that the energy barrier to insertion was $\leq$ 1 kcal mol^{-1} [4], this provides experimental support for the higher values of heat of formation at room temperature. This suggestion has been supported by *ab initio* calculations carried out by Gordon and Gano[7]. The calculations for the potential energy surface of the silylene reaction with hydrogen gave an activation energy of 1.7 kcal mol^{-1}.

Subsequently Baggott *et al.*[8] extended Jasinski's room temperature study to cover a range of 298-333K and found k to be in good agreement with Jasinski's reported value[4,5]. The authors found an almost temperature independent rate constants of $1.88x10^{-12}$ $cm^3molecule^{-1}s^{-1}$. The uncertainties in the rate constants were *ca.* 7%. This indicated that any activation energy was very small and certainly <1.57 kcal mol^{-1}. A slightly negative activation energy cannot be ruled out.

In this work we report the first kinetic studies for the reaction of SiD_2 + D_2 at a wide range of pressures and temperatures. Kinetic studies of SiD_2 are of particular interest in two respects, *viz*
(i) a comparison with SiH_2 data will help elucidate reaction mechanisms,
(ii) many SiH_2 reactions exhibit pressure dependence often making it problematical to obtain high pressure

* Corresponding author: 100108@uotechnology.edu.iq

limiting rate constants. Isotopic scrambling effects make it possible to circumvent this problem with SiD_2 studies.

2 SiD_2 precursor and visible absorption spectrum

$PhSiD_3$ was prepared by reduction of phenyltrichlorosilane ($PhSiCl_3$) with $LiAlD_4$ in dry ether [3]. The prepared $PhSiD_3$ was purified by distillation and found to be 99.8% pure (isotopic purity >92.6%). The $PhSiD_3$ precursor was chosen because its analogue phenylsilane, $PhSiH_3$ proved to be a good precursor for SiH_2. Although SiH_2 formation is not the major product of $PhSiH_3$ photolysis[9], nevertheless, its shown observation, suggest that SiD_2 should be readily detected from $PhSiD_3$. The high quality decay signals observed in the current work reveals that this is true and that if other decomposition pathways (*eg* to $PhSiD + D_2$) are occurring, they do not effect measurements of the rate constants for SiD_2. A typical averaged decay trace is shown in figure 1.

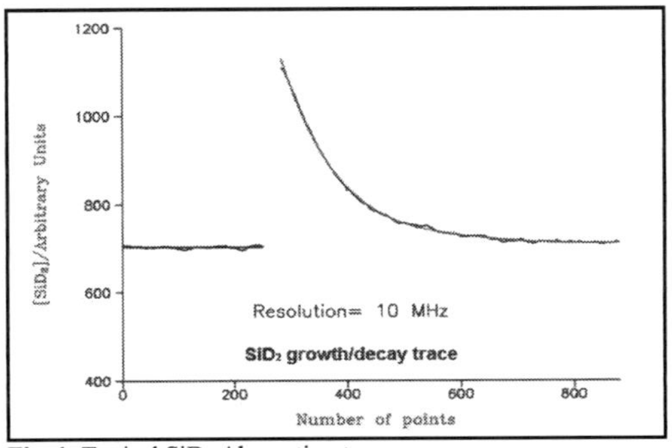

Fig. 1. Typical SiD_2 Absorption trace
(3 mTorr of $PhSiD_3$, 4 Torr D_2 and P_T= 5 Torr in SF_6).

The absorption spectrum of SiD_2, formed from the flash photolysis of a mixture of SiD_4 and D_2, was first investigated by Dubois *et al.* in the region 16175-19200 cm^{-1} [10]. They recorded a visible spectrum of the $(^1B_1)\leftarrow(^1A_1)$ band and found the bending vibrational frequency in the excited state to be 610 nm. However a more detailed spectroscopic study of SiD_2 radicals was reported by other researchers such as Fukushima *et al.* and Muramoto *et al.* [11-13]. Fukushima *et al.* [11] photolysed $PhSiD_3$ at 193nm in the supersonic free jet and observed 11 vibronic bands of the SiD_2 A $(^1B_1) \leftarrow$ X $(^1A_1)$ transition in the wavelength region between 21739.13 and 15625 cm^{-1}.

Because little is known about the SiD_2 radical, any fresh spectroscopic data (as well as kinetic data) are of practical and theoretical interest. Laser spectroscopic investigation of SiD_2 at higher resolution than earlier studies has reached similar conclusions in our lab [3]. Mason *et al.*[3] recorded the absorption spectrum of SiD_2 in the gas phase, using the flash photolysis kinetic absorption technique. SiD_2 was generated by photolysing 10mTorr of $PhSiD_3$ in 5Torr of SF_6 using a series of one-shot photolysis experiments. The authors observed an absorption spectrum in the 17384-17391 cm^{-1} region (the laser linewidth was approximately $3x10^{-5}$ cm^{-1}) with a strong band at 17387.07cm^{-1}, which was attributed to an unknown rotational line $A(^1B_1)(0,3,0)\leftarrow X(^1A_1)(0,0,0)$ vibronic transition of SiD_2. This assignment was based on the original work of Fukushima *et al.*[11]. A 6 cm^{-1} composite spectrum of vibronic transition of SiD_2 is shown in figure 2 (a). Figure 2 (b) also shows the absorption spectrum of SiH_2 for comparison reason. Approximately 8 peaks were displayed for SiD_2 absorption spectrum. The 17387.07 cm^{-1} was found to be the most intense peak. Therefore this peak was chosen to observe the SiD_2 species in this kinetic study. It has been estimated that the intensity of this peak is approximately a quarter of that for the transition used to monitor the SiH_2 species in kinetic studies under similar conditions[3].

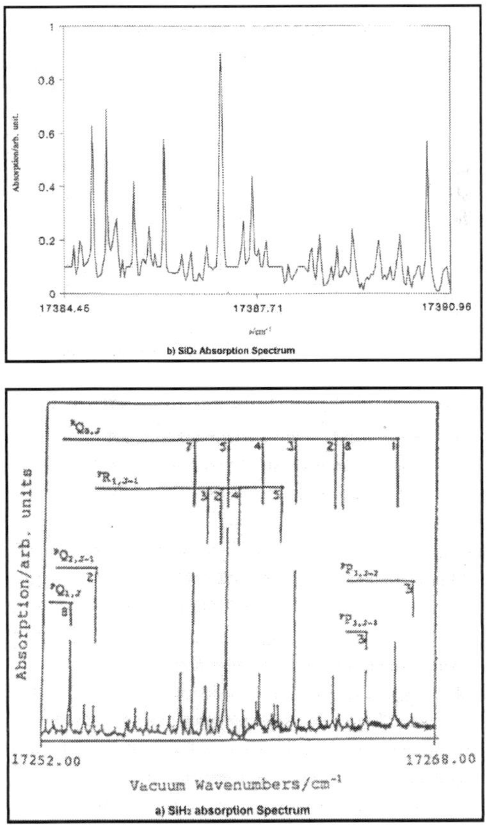

Fig. 2. Absorption spectrum of both a) SiH_2 [4] and b) SiD_2[3].

3 Experimental Setup

Laser flash photolysis/laser absorption apparatus is diagrammatically represented in figure bellow (figure 3).

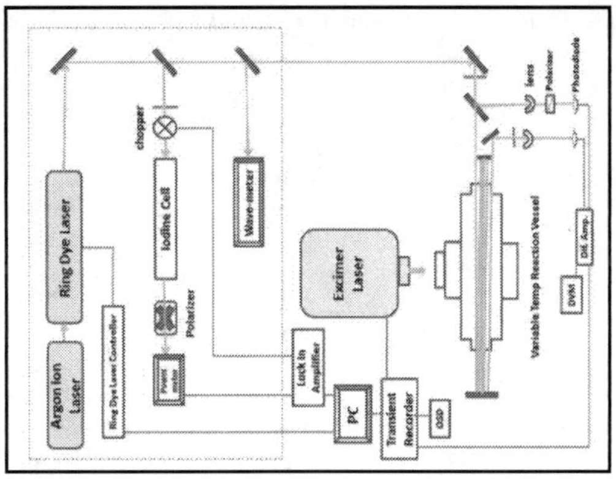

Fig. 3. Schematic diagram of the experimental apparatus.

A mercury-free, glass vacuum line was connected to purpose built variable photolysis cell fitted with a high quality crown glass end windows at Brewster's angle backed by a rotary pump. SiD_2 was produced by the 193 nm flash photolysis of phenylsilane ($PhSiD_3$) using an Oxford KX2, ArF excimer laser. The emitted pulses have 0.1-1J of energy delivered in a 10 ns light pulse with a repetition rate of 1 Hz. SiD_2 concentrations were monitored in real time by means of a Coherent 699–21 single-mode dye laser pumped by an Innova 90–5 Ar^+ laser using Rhodamine 6G in solution as an active medium, which provides tunable radiation in the range 560-610nm. The monitoring laser beam was multipassed between 24-48 times through the reaction zone to give an effective path length of between 0.6-6 m. The monitoring laser was tuned to 17387.07 cm^{-1}. A differential amplifier was connected to two photodiodes and then fed into Datalab Dl910 transient recorder (20 MHz resolution). The transient recorder's external trigger was activated by radiofrequency noise produced when the excimer laser was fired. Traces were observed during runs and interfaced to a computer via a commercial interface (Camplus) where the signal could be displayed and analyzed.

4 Results

The SiD_2 exponential decay curves were recorded and averaged for 10 excimer laser shots. The degree of depletion of the photolysis mixture was negligible. Experiments were carried out with gas mixtures containing a few mTorr of $PhSiD_3$, varying quantities of D_2 up to 9.5 Torr, and inert diluent bath gas (SF_6) at total pressures from 5 to 100 Torr to test for pressure dependence. For each pressure, a series of runs (7 points) was performed at different substrate concentrations, as been illustrated in figure 4.

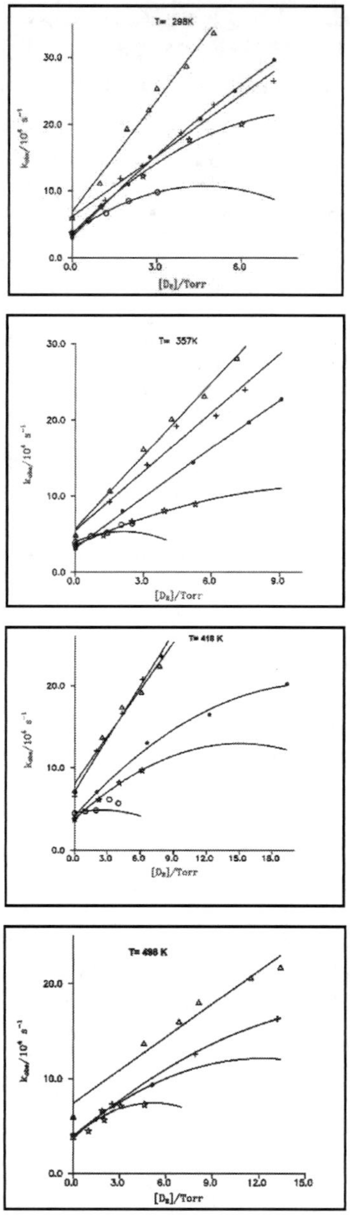

Fig. 4. Second-order plots for the reaction of SiD_2 + D_2 at various total pressures (in SF_6) and Temperatures.
O, 5; ★, 10; •, 20; +, 50; Δ, 100 Torr.

Only second order plots for data recorded at a total pressure above 20 Torr in SF_6 were linear. Plots for total pressures below this were curved. The quadratic procedure using the polynomial regression was applied for these cases in order to obtain the second order rate constants as gradients of the plots. The intercept corresponds to loss of silylene in the absence of substrate i.e. by reaction with $PhSiD_3$. The correlation of fitting constants means that the rate constants are somewhat less reliable by this procedure.
Second order rate constants were calculated using the following expression:

$$k(cm^3\ \text{molecule}^{-1}s^{-1}) = \frac{Gradient(Torr^{-1}\ s^{-1}) \times T(K)}{F}$$

The conversion factor, F, was obtained from the ideal gas equation (PV=nRT where R=8.3145 J $K^{-1}mol^{-1}$). Its value is F= 9.657×10^{18}.

The second order rate constants for these experiments, representing over 100 measurements of individual rate constants are shown in figure 5 and listed in table 1.

The uncertainties in individual second order rate constants are about *ca.* ± 20% for total pressures <20 Torr and ± 10% for total pressure >20 Torr, but are not shown in the figures to avoid complicating them.

Table.1 Absolute rate constants for the reaction of $SiD_2 + D_2$

P/Torr	$k/10^{-12}\ cm^3molecule^{-1}s^{-1}$			
	298K	357K	418K	498K
5	0.91±0.1	0.45±0.1	0.23±0.1	-
10	1.33±0.1	0.53±0.1	0.53±0.1	-
20	1.35±0.1	0.83±0.1	0.63±0.1	0.69±0.2
50	1.21±0.2	1.16±0.2	1.04±0.1	0.66±0.2
100	1.72±0.2	1.36±0.2	1.00±0.1	0.78±0.2

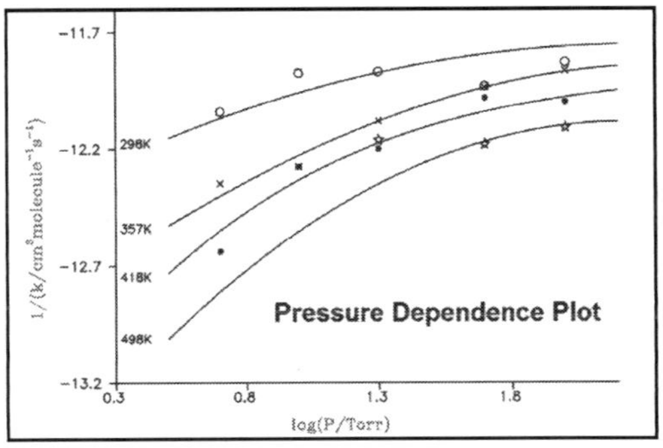

Fig.5. Pressure dependence of second-order rate constants for $SiD_2 + D_2$ at different temperatures in SF_6 buffer gas.

Comparison of absolute rate constants can only be made at 300K between the reactions of SiH_2+H_2 and SiD_2+D_2 due to big differences of temperatures used in these studies, see figure 6. The graph shows that the SiH_2+ H_2 reaction is approximately ca. 60% faster than SiD_2+D_2 reaction at high pressure.

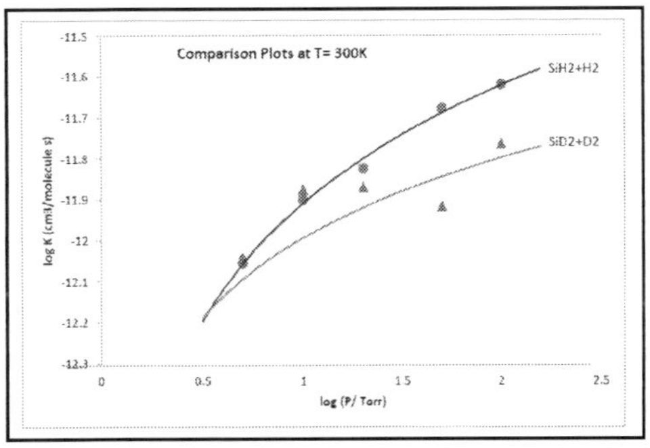

Fig.6. Pressure dependence of second-order rate constants for SiH_2+H_2 and SiD_2+D_2 at T=300K.

The measured rate constants demonstrate the reaction is relatively slow and rate constants are weakly pressure dependent at all temperatures, consistent with a third-body mediated association process. In addition, SiH_2+ H_2 reaction is approximately ca. 60-70% faster than SiD_2+D_2 reaction at higher total pressure.

5 Extrapolation by empirical procedures

Since SiD_2+D_2 study lies in the fall-off region at accessible pressures, a reliable method of extrapolation to infinite pressure must be employed. A number of methods have been suggested and used by various workers.

The earliest extrapolation procedure employed plots of k^{-1} versus P^{-1}. This plot follows from the Hinshelwood-Lindemann model. Recently Oref and Rabinovitch[14] have suggested another empirical procedure which theoretically should yield more accurate infinite pressure rate constants since the precise nature of the extrapolation is determined by the actual data. They suggest a plot of k^{-1} versus P^{α} where α lies between 0 and 1 and is optimised to give the best linear plot. Values of α on either side of the optimum value will give curved plots which would lead to an overestimate and underestimate of k^{∞} resp. It is important to state that there is no absolute theoretical justification for either of the methods described above. They are simply empirical procedures which have been found to give reasonable linear plots. In the present work the latter method of extrapolation (where α= 0.5 and 1) has been tried and the results are compared.

When data are only available at pressures well into the fall off region it becomes clear that the choice of method of extrapolation becomes crucial, since different methods give different answers. In the current work it was extremely difficult to judge the quality of the fits because of data scatter, values of k^{∞} (represented by $(\text{intercept})^{-1}$) had a large uncertainty. Therefore a modified procedure was adopted here in which a number of points were selected from interpolated curves (drawn by hand). These interpolated fall-off curves were used to represent the data in order to obtain k^{∞} with minimum error (see figure 7).

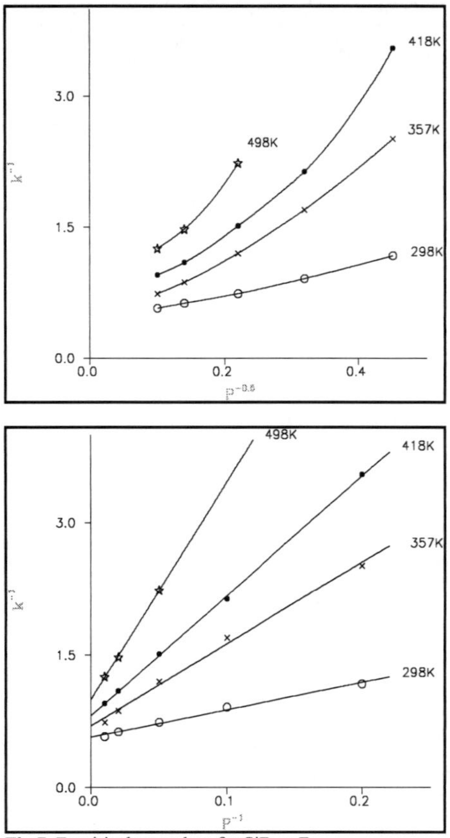

Fig.7. Empirical procedure for $SiD_2 + D_2$.

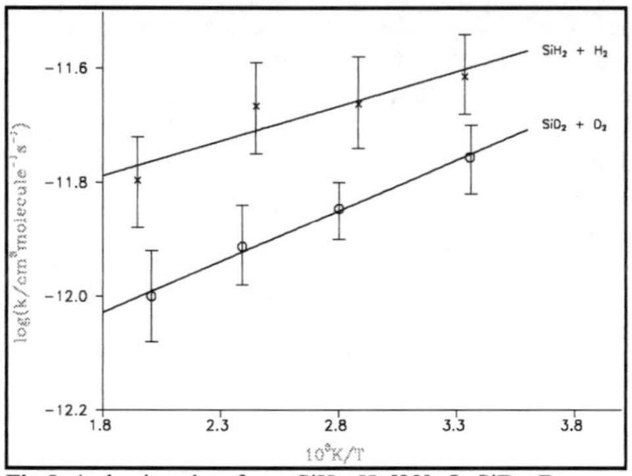

Fig.8. Arrhenius plots for x, SiH_2+H_2 [23]; O, SiD_2+D_2.

We find that the plot is not absolutely linear. Consequently some error in the infinite pressure rate constants obtained using a least-squares line must be expected. If we consider the k^{-1} versus $P^{-0.5}$ we find that such plots suffer from marked curvature making extrapolation difficult and leading to an underestimated of k^{∞}. On the other hand the k^{-1} versus P^{-1} plots are much less curved. It must be noted, however, that scatter in the experimental data can make calculation of an optimum α difficult and may limit the applicability of this method.

The temperature dependence for the reaction may be expressed in Arrhenius form. The derived Arrhenius parameters are shown in table 5. Figure 8 shows a graph of the Arrhenius plot for both SiH_2+H_2 [15] and SiD_2+D_2 reactions. These data show a number of interesting features. First the reaction has a negative activation energies E= - 0.82±0.05 kcal mol^{-1}. Secondly there is a significant difference between SiH_2+H_2 and SiD_2+D_2. The former reaction is ca. 60% faster than the latter.

6 RRKM modelling of experiments

In order to try to fit the observed pressure dependence of the insertion of SiD_2 + D_2, RRKM calculations were carried out on the pressure dependence of the inverse silane, SiD_4 decomposition in the conventional manner.

$<\Delta E>_{down}$ were based on the assumption of a fixed energy removed on each collision (stepladder model). However we did not include some other correction factors (eg F_{rot}) which could have some effect. For a small molecule like silane (highly excited), the anharmonicity effect was found to be important[16,17]. This correction factor influences the density of states of the reactant at an internal energy equal to the critical energy. The low pressure limiting rate constant is proportional to the anharmonicity factor, by increasing the value of this factor better agreement with the data was obtained [17-19]. It has been suggested that increasing the value of this factor has roughly the same effect as increasing the "looseness" of the transition state.

The fall-off predictions are very sensitive to some of the parameters and not very sensitive to others. The exact assignments of reactant frequencies, transition state frequencies, and reactant path degeneracy are not very important as long as these assignments reproduce the required high pressure A-factor. On the other hand, assignment of bath gas collision efficiencies and their temperature dependencies can be very important, particularly when the reaction is well into its pressure-dependent region and the bath gas molecules are weak colliders.

The effect of varying the energy level grain size (GRAIN) is rather unpredictable and its choice affects the position of the fall-off curves for both strong and weak collision situations. This parameter is used to define the energy gap in the quasi-quantised energy scale covering the energy range of interest. It is clear that the smaller its value, the more closely packed are the levels. Therefore in principle smaller grain sizes are best (although this increases the size of programme storage required and decreases the speed of calculations). However the calculation is also sensitive to the positioning of the critical energy, E_o with the quantised energy levels. The best situation appears to be when E_o

lies in the center of a grain. The choice of minimum energy value, E_{min} and grain size, determines this. In practice there were chosen to get closer to this condition. The energy range starts from just below the initial energy and goes up to 2-3 times its value. In the present calculation for silylene reaction with hydrogen this parameter was set to be 50 cm^{-1}. The vibration frequencies assigned to the SiD_4 complex were based on those from ab initio calculations of Gordon et al.[20]. The details of the model and parameters used are shown in tables 2, and 3.

Table 2. RRKM parameters for SiD_4 decomposition model, predicted by Gordon *et al.*[20]

	SiD_4	SiD_4^+
	1597(3) 1558(1) 701(2) 681(3)	1579(1) 1551(1) 1488(1) 1163(1) 721(1) 681(1) 528(1) 498(1)
I_i/amu A^2	11.777 11.777 11.776	13.685 11.558 9.707
Path degeneracy	12	
Reaction coordinate	681	
E_o/ kcal mol^{-1}	56.18 (298K)	
log(A/s^{-1})	14.38 (298K)	
Z_{LJ}/ 10^{-10} cm^3 molec^{-1} s^{-1}	4.354 (SF_6, 298K)	

Table 3. Z_{LJ} for SiD_4 decomposition.

Temperature/K	$Z_{LJ}/10^{-10}$ cm^{3}molecule^{-1}s^{-1}
298	4.354
357	4.487
418	4.620
498	4.785

The A-factors used here are those obtained using the vibration wavenumbers for both the molecule and the Transition State (employing Smol programme) [15]. Since both SiH_4 and SiD_4 have the same electronic energies, the critical energies can be obtained from the standard formula ($E_o=E_{el.}+\Delta zpe$). Δzpe is defined as the difference between the zero-point energies of the reactant and activated complex. The activation energies then can be calculated using the equation $E_a = E_o + RT + \langle E^+\rangle - \langle E\rangle$. The results of these calculations are shown in table 4.

Table 4. Activation energies and A-factors for SiD_4 decomposition.

T/K	log(A/s^{-1})	E_a/kcal mol^{-1}	E_o/kcal mol^{-1}
298	14.38	56.83	56.18
357	14.45	56.80	56.04
418	14.51	56.74	55.87
498	14.57	56.64	55.65

The calculations were carried out using both a strong and a weak collision model to generate fall-off curves at four different temperatures. A weak collision ($\langle\Delta E\rangle_{down}$= 2.3 kcal mol^{-1}) was fixed to fit the rate constants at all four temperatures and is reasonably consistent with collisional efficiencies found with similar systems[20]. However these calculations were carried out without including the F_{an} factor which was assumed to have a small effect.

The results are plotted in figure 9 with experimentally determined rate constants. The high pressure limit rate constants were derived from the isotope effect calculations combined with SiH_2 + D_2 rate data[15]. Reasonable agreement was obtained with the data when compared to that of the previous reaction, except for T=357K where the theoretical fit and data were in disagreement. However there are probably other choices of parameters that would have produced equally good fits.

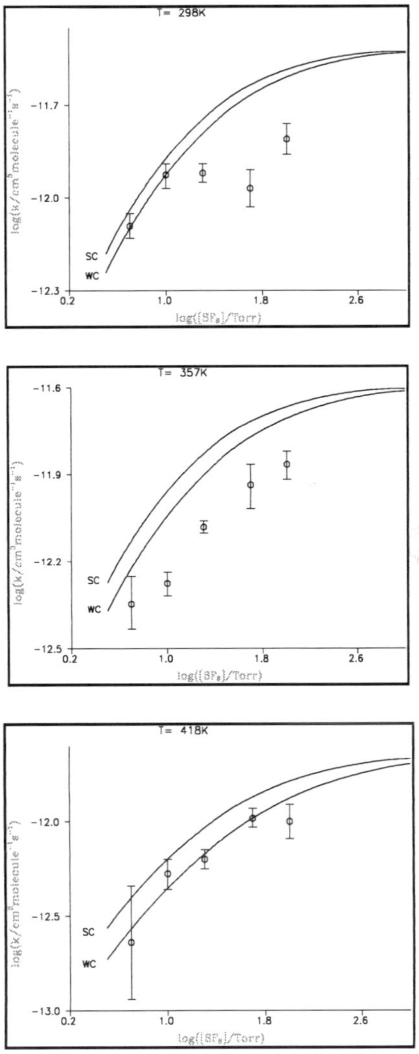

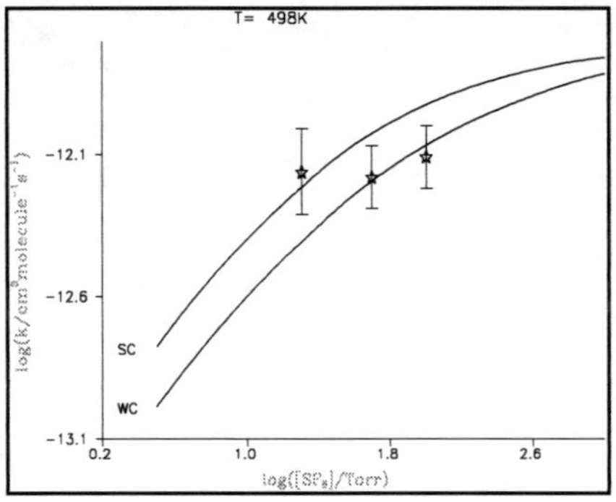

Fig.9. Pressure dependence of second-order rate constants for $SiD_2 + D_2$ at different temperatures in SF_6 buffer gas (solid lines are RRKM fits). sc, Strong collision; wc, Weak collision

7 Discussion

These studies proved to be more difficult than earlier ones for silylene with olefins[21]. This arose because the rate constants were smaller and a significant amount of hydrogen was required to observe a reaction. This meant that at low total pressures, reaction of silylene with precursor was always a substantial fraction of the total decay. The magnitude of the intercept of the second order plots were kept as small as possible. This was done by reducing the partial pressure of SiD_4 to the minimum consistent with signal to noise constraints.

This restricted the range of possible substrate pressures and temperatures over which reactions could be studied. At high partial pressures of hydrogen necessary to obtain reaction, SF_6 is present in insufficient quantity to ensure that it is the dominant bath gas collider. This causes curvature in the second order plots (at fixed total pressures). By using a polynomial procedure (quadratic), we are assuming that the slope at $[D_2/Torr]=0$ will provide a reliable estimate of the first-order rate coefficients, k_{obs}, for the removal of SiD_2 in the presence of hydrogen. However this leads to higher uncertainties in this case compared to those of other studies (*eg* $SiH_2 + C_2H_4$)[21].

A high pressure limit rate comparison is shown in table 5 using different extrapolation procedures.

Table 5. High pressure limit rate comparison for the reaction of $SiD_2 + D_2$ using three different procedures.

T/K	$k^{\infty}/10^{-12}$ cm^3 molecule^{-1} s^{-1}		
	RRKM	**Empirical Procedure**	**Isotope Effect**
298	2.24	1.75 ± 0.35	3.38
357	1.78	1.43 ± 0.29	2.54
418	1.58	1.22 ± 0.24	2.23
498	1.41	1.00 ± 0.20	1.92

A comparison is also made for the activation energies and the A-factors for silylene reaction with hydrogen and its isotopic analogues in table 6 (obtained from extrapolation of experimental data to the high pressure limit using empirical procedure).

Table 6. Arrhenius parameters for SiH_2+H_2 and isotopic variants

Reaction	log (A/ cm^{3}molec^{-1}s^{-1})	Activation energy kJ mol^{-1}
$SiH_2 + H_2$	-12.01[a] -11.75[b]	-2.34 -1.84
$SiD_2 + D_2$	-12.35[a] -12.09[b]	-3.41 -3.52
$SiH_2 + D_2$	-12.07[c]	-2.03
$SiD_2 + H_2$	-11.87[d]	-2.54

[a]. Obtained by extrapolation (empirical procedure) [15]
[b]. Based on isotope effect calculations[15]
[c]. P_T= 10 Torr[15], [d]. P_T= 10 Torr[3]

The following figures can be drawn from this table. Arrhenius parameters of E_a= -0.48 ± 0.33 kcal mol^{-1} and A-factor= 0.80±0.35×10^{-12} cm^{3}molecule^{-1}s^{-1} can be averaged. The activation energy for the reaction (SiH_2+H_2) is approximately a factor of 1.5 smaller than that obtained from reaction (SiD_2+D_2) by extrapolation to the high pressure limit.

A reasonable agreement was seen between theory and experiment for SiD_4 decomposition. However there are probably other choices of parameters that would have produced equally good fits. Parameters such as the average energy removal, $\langle\Delta E\rangle_{down}$, and the high pressure limit rate constant, k^{∞}, can be adjusted to obtain a better fit (see table 5).

An example is shown in figure 10 at T= 357K, which shows the fall-off curve when $\langle\Delta E\rangle_{down}$= 2.3 kcal mol^{-1} and k^{∞}= 1.78 x 10^{-12} cm^{3}molecule^{-1}s^{-1}.

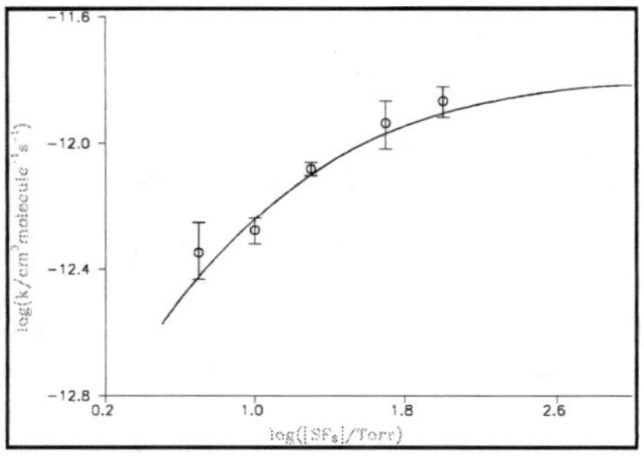

Fig. 10. Pressure dependence of second-order rate constants for $SiD_2 + D_2$ at T= 357K (solid line is RRKM fit)

8 Summary and future work

Time-resolved study of SiD_2 generated by laser flash photolysis of ($PhSiD_3$) has been carried out to obtain rate constants for its bimolecular reactions with D_2. The reaction was studied over the pressure range 5-100 Torr (in SF_6 bath gas) at four temperatures in the range 298-498K. The reaction was found to be pressure dependent at all temperatures, consistent with a third-body mediated association process. The reactions were modelled using empirical and theoretical procedures to obtain the high pressure limit rate constants. The empirical procedure employed the Hinshelwood-Lindemann model. The data obtained using these extrapolations were fitted to the Arrhenius equation. RRKM modelling, based on the transition states (SiD_4) of Gordon *et al*, gave a reasonable fit to the experiments. Further work is required in the following areas; (i) the RRKM calculation described in this work can be improved by considering some other correction factors in addition to that of F_{an}. (ii) the pressure dependence of the pyrolysis of silane also needs to be reinvestigated over a wider pressure range in order to obtain an improved value for k^{∞} for $SiD_4 \rightarrow SiD_2 + D_2$. This would then improve the value for the heat of formation of silylene.

References

1. N Al-Rubaiey, HM Frey, BP Mason, C McMahon, R Walsh, Chem. Phys. Lett., **203**, (3-4), 301-305 (1993).
2. N Al-Rubaiey, IW Carpenter, R Walsh, R Becerra, Mark S Gordon, The Journal of Phys. Chem. B, **102** (44), 8564-8572 (1998).
3. B.P Mason, H.M. Frey, and R. Walsh, J. Chem. Soc. Faraday Trans., **89**, 4405 (1993).
4. J.M. Jasinski, J. Phys. Chem., , **90**, 555 (1986).
5. J.M. Jasinski, J.O. Chu, in "Silicon Chemistry", ed. by J.Y. Corey, E.R. Corey, and P.P. Gaspar, Ellis Horwood, Chichester, UK, (1988).
6. J.M. Jasinski, J.O. Chu, J. Chem. Phys., **88**, 1678 (1988).
7. M.S. Gordon, D.R. Gano, J.S. Binkley, M.J. Frisch, J. Am. Chem. Soc., **108**, 2191 (1986).
8. J.E. Baggott, H.M. Frey, K.D. King, P.D. Lightfoot, R. Walsh, I.M. Watts, J. Phys. Chem., **92**, 4025 (1988).
9. J.E. Baggott, H.M. Frey, P.D. Lightfoot, R. Walsh, Chem. Phys. Lett, **22**, 125 (1986).
10. I. Dubois, G. Herzberg, R. D. Verma, J. Chem. Phys., **47**, 4262 (1967).
11. M. Fukushima, S. Mayama, K. Obi, J. Chem. Phys., **96**, 44 (1992).
12. Fukushima, M.; Obi, K., , J. Chem. Phys., **100**, 9, 6221 (1994)
13. Muramoto, Y.; Ishikawa, H., J. Chem. Phys., **122**, 15, 154302 (2005).
14. I. Oref and B.S. Rabinovitch, J. Phys. Chem., **72** (13), 4488 (1968).
15. N. Al-Rubaye and R. Walsh, to be published.
16. J. Troe, J. Phys. Chem., **83**, 114 (1979).
17. K.F. Roenigk, K.F. Jensen, and R.W. Carr, J. Phys. Chem., **91**, 5736 (1987).
18. H.K. Moffat, K.F. Jensen, R.W. Carr, J. Phys. Chem., **95**, 145 (1991).
19. J.M. Jasinski and J.O. Chu, J. Chem. Phys., **88**, 71678 (1988).
20. M.S. Gordon, private communications to R. Walsh.
21. N. Al-Rubaiey and R. Walsh, J. Phys. Chem., **98** (20), 5303 (1994).

Microstructure-mediated Optical Effects in Southern African Snakes

Ishan Singh[1,*], and *Graham* Alexander[1]

[1]School of Animal, Plant, and Environmental Sciences, University of Witwatersrand, P.O. Wits, 2050, South Africa

Abstract. The scales of the African Viper *Bitis arietans* were tested for optical effects. Spectral intensity was recorded at incident angles over the visible spectrum for dark, pale, and ventral scale regions. The lowest spectral intensity recordings were associated with scales which have the greatest level of micro-structuring. Our results indicate that scale appearance in *B. arietans* is a product of microstructure-mediated optical effects. The optical effect may play a role in improving the ecological performance of the snake in its natural environment.

1 Introduction

Some species of snake have microstructures called ‘microornamentation’ on their scales [1]. The spatial arrangement of microornamentation determines the optical properties of the scale, typically its reflectance. Optical effects such as structural colour are produced through the interaction of electromagnetic radiation with microornamentation [2]. The resulting light reflected off an animal's integument determines the animal's visual appearance [3] and thus impacts its ecological relationships. This study aims to investigate if microornamentation creates optical effects in the scales of *B. arietans*.

2 Methods and Materials

Shed skin from a *B. arietans* was collected from the National Zoological Gardens (Pretoria, Reference: NZG/RES/P/001/F/02 P15/03. Reference: 17-04-2015-O). We sampled scales from five areas along the length of the body of each specimen. Scale colour was recognised using spectrophotometry.

For microscopy, samples sputter coated with ± 10 nm of Carbon and ± 5 nm Chromium were observed at an accelerating voltage of 30.00 kV using a FEI Nova 600 Nanolab FIB/ SEM (FEI, Munich, Germany). The mean height and mean density of microstructures were calculated for 10 structures per sample and for five

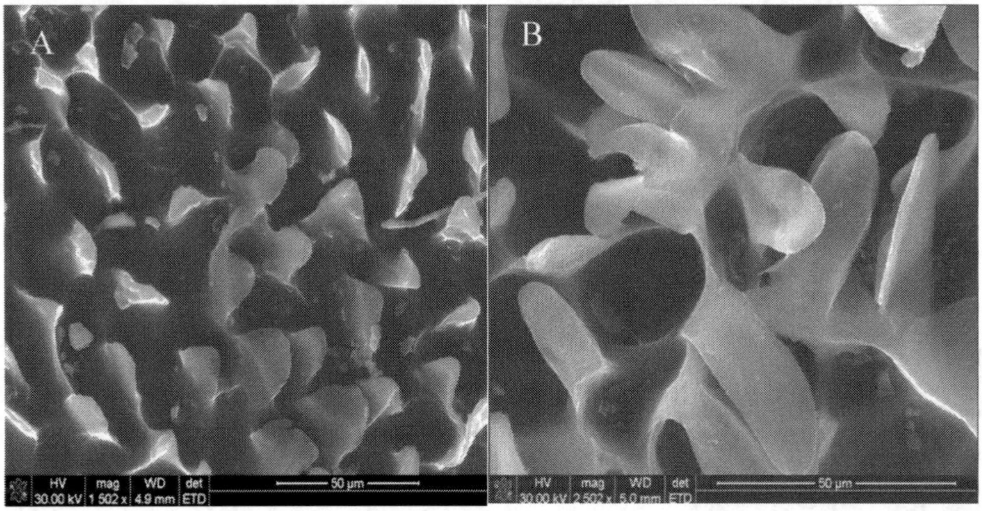

Figure 1. Electron micrographs of pale (A) and dark (B) scale regions of *B. arietans* showing microstructures on the scale surface. Scale bars: 50 um

* Corresponding author: ishp2singh@gmail.com

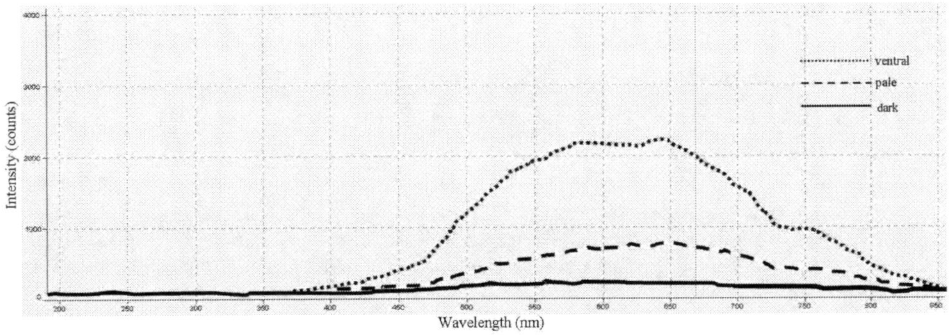

Figure 2. Spectra recorded for pale, dark, and ventral scales at incident angles over the visible spectrum.

samples per individual. Structure density per sample was counted using a haemocytometer in five 100 mm x 100 mm quadrats.

To measure optical effects, twenty 1 cm^2 pieces of skin were excised from exuviam. Of these, 10 samples were sputter-coated with 15 nm Gold-Palladium to control for pigmentary differences, and 10 samples were left uncoated. We took spectral readings (Ocean Optics Inc, Dunedin, Florida, USA) in triplicate over the visible spectrum at incident angles.

3 Results

Microstructures were observed on the surface of both dark and pale scales (fig. 1), but not ventral (belly) scales. A morphometric analysis indicated that the density of microstructures on dark scales (mean density = 2696 ± 613 mm^{-2}) and pale scales (mean density = 2640 ± 497mm^{-2}) did not differ significantly ($t_{8,0.8} = 0.26$; $p = 0.80$). There is a significant difference in height of microornamentation (Mann-Whitney U-statistic = 184.00; $p < 0.0001$) between dark (mean height = 40.83 ± 8.33 μm) and pale (mean height = 17.29 ± 4.05 μm) scale regions.

The recorded spectra of dark, pale, and ventral regions differed greatly (fig. 2). Dark regions had the lowest recorded intensity of all, while both dark and pale regions were lower than the ventral regions. Ventral scales, due to their lack of microstructures, were used as a control.

4 Conclusion

Our results show that colour differences correspond to the height of scale microstructure. We conclude that scale colour in *B. arietans* is a product of optical effects created by microornamentation. *Bitis arietans* is widely distributed across Africa [4], and as an ambush forager may remain in ambush positions for extensive periods of time. Thus, there is great reliance on visual and chemical camouflage to avoid detection by predators and prey [5]. We suggest that the optical effects created by microornamentation contribute to the snake's visual camouflage, and thus has an important function in the ecology of this species.

The authors wish to acknowledge the National Research Foundation, the National Zoological Gardens, and The University of Witwatersrand, South Africa.

References

1. M. Spinner, A. Kovalev, S. Gorb, G. Westhoff. Nature **3** (2013)
2. P. Vukusic. Ophthalmic Physiol Opt **30** (2010)
3. P. Vukusic, D. Stavenga. J. R. Soc. Interface **6** (2009)
4. G.J. Alexander, J. Marais. *A guide to the reptiles of southern Africa* (Struik Nature, Cape Town 2007)
5. A.K. Miller, B. Maritz, S. McKay, X. Glaudas, G.J. Alexander. Proc. R. Soc. B **282** (2015)

EPJ Web of Conferences **139**, 00006 (2017) DOI: 10.1051/epjconf/201713900006
NANOP 2016

Sensitive Detection of Biomolecules by Surface Enhanced Raman Scattering using Plant Leaves as Natural Substrates

Vipul Sharma and Venkata Krishnan[*]

School of Basic Sciences and Advanced Materials Research Center, Indian Institute of Technology Mandi, Kamand, Mandi 175005, H.P., India

Abstract. Detection of biomolecules is highly important for biomedical and other biological applications. Although several methods exist for the detection of biomolecules, surface enhanced Raman scattering (SERS) has a unique role in greatly enhancing the sensitivity. In this work, we have demonstrated the use of natural plant leaves as facile, low cost and eco-friendly SERS substrates for the sensitive detection of biomolecules. Specifically, we have investigated the influence of surface topography of five different plant leaf based substrates, deposited with Au, on the SERS performance by using L-cysteine as a model biomolecule. In addition, we have also compared the effect of sputter deposition of Au thin film with drop-cast deposition of Au nanoparticles on the leaf substrates. Our results indicate that L-cysteine could be detected with high sensitivity using these plant leaf based substrates and the leaf possessing hierarchical micro/nanostructures on its surface shows higher SERS enhancement compared to a leaf having a near-planar surface. Furthermore, leaves with drop-casted Au nanoparticle clusters performed better than the leaves sputter deposited with a thin Au film.

1 Introduction

There are various types of important biomolecules that are responsible for the maintenance of metabolic processes in living organisms [1]. In the scientific field, the detection of these biomolecules is essential in various analytical, medical, biochemical and pharmaceutical applications [2,3]. Biomolecules are generally composed of organic compounds, which include large macromolecules, like proteins, carbohydrates, lipid s, and nucleic acids, as well as small molecules such as various metabolites and natural products [3]. In general, the most common techniques used to detect and study these biomolecules include polymerase chain reaction (PCR) and enzyme linked immunsorbent assay (ELISA) [4]. The other methods to the study the biomolecules include various microscopic and spectroscopic techniques, such as fluorescence microscopy, Fourier transform infrared spectroscopy, UV-vis absorption spectroscopy and Raman spectroscopy [5]. All the methods of detection and analysis of biomolecules are unique and give valuable insights about the structure and composition of biomolecules. Amino acids are considered as the basic building blocks that combine together to form proteins. L-cysteine, is reported as a 'non-essential' amino acid that is typically synthesized by the human body and is responsible for the stabilization of secondary structure of proteins [6]. Since this molecule plays an important role in deactivating free radicals (antioxidant), it has been the focus of several studies [7].

Raman spectroscopy is a valuable investigation tool that employs distinctive vibrational patterns for the detection of molecules [8]. It has direct application in the chemical and bio sensing, however its use is very much limited due to low intensity of normal Raman scattering signals [9]. Nevertheless, much more sensitive Raman techniques are being developed such as, surface enhanced Raman scattering (SERS), wherein the method provides valuable information on the molecules and their local interactions than normal Raman spectroscopy by leveraging the characteristic and exclusive properties of plasmonic nanoparticles leading to enhanced detection limits while maintaining the very good sensitivity [10]. When the electromagnetic radiation interacts with the oscillating cloud of conduction electrons of plasmonic nanoparticles, it leads to the electromagnetic energy confinement around the nanoparticles. Significant Raman signal amplification can be achieved by the introduction of a Raman active molecule in close proximity to the nanoparticles, and within the confined electromagnetic field [10,11]. The Raman scattering from an analyte adsorbed on the nanostructured metal surface can significantly increase in comparison to solution. The phenomenon was first observed by Fleischman in 1974 and elaborated by Van Duyne in 1977 [12]. The nanostructures of noble and transition metals onto which the molecules are adsorbed can lead to 10^{10} to 10^{15} fold enhancement of the intensity of Raman scattered light and the effect is reported to be strongest on Ag followed by Au and Cu [10].

Further moving in this field, improvement in the sensitivity of SERS signal and performance has been amplified by arranging the nanostructures into a three-dimensional structure by many ways such as placing the plasmonic nanostructures directly or combining plasmonic nanostructures with non-plasmonic frameworks [13]. The preparation of 3D SERS substrates by merging plasmonic nanostructures with a roughened or grooved substrate is generally done by decorating noble metal nanostructures on specifically designed 3D frameworks [14]. In some of the previously reported

[*] Corresponding author: vkn@iitmandi.ac.in

DOI: 10.1051/epjconf/201713900006

works, 3D SERS substrates have been produced by decoration of noble metal nanoparticles on various 3D templates [15,16]. For the creation of 3D nanostructured templates, methods that allow ordered hierarchy in the structure, like Langmuir-Blodgett technique [17], electrochemical methods [18], and lithography methods [19] are normally employed. The fabricated substrates exhibit high density of the metal nanostructures ordered in clusters. These interact with analytes and abundant nanometer level gaps to dramatically enhance the local electromagnetic fields upon excitation by incident laser. In the recent years, biodegradable substrates, such as biological fibers have been used to create 3D templates for decoration of plasmonic nanostructures [20]. There are also some reports where paper made up of cellulose fibers has been employed to create the 3D templates for the SERS applications. Very good SERS activity is reported for traditional SERS supports and many engineered substrates with improved designs, but they do have some limitations as it is very difficult to assemble the nanoparticles in a well-ordered and controllable manner for efficient and sensitive SERS detection [21].

Plasmonic metal decorated natural and bioinspired surfaces have already gained attention for SERS applications, given their good plasmonic properties and 3D hierarchical structure for the fabrication of the efficient SERS substrates [22]. There are natural surfaces present in different flora and fauna, which possess vibrant hierarchical structures, wettability and adhesion related properties [22]. For SERS applications, there are reports where researchers have used butterfly wings as a substrate, which provided very good enhancement factors and very low detection limits as well [23]. Recent reports include rose petals and Taro leaf which were directly used as 3D templates for SERS where the nanoparticle clusters were very nicely arranged over the microstructures present on the surfaces and provided very good SERS activity [24]. In one of our recent works, we have reported on the use of gold deposited plant leaves for SERS application for the detection of polar and non-polar analytes, wherein we had elaborated the role of surface morphology, wettability and deposition technique in determining the enhancement factor and detection limit [25]. The performance of the SERS substrates is generally dependent on two factors: the concentration of analytes and the existence of strong electromagnetic field [15,26]. Usually the Raman signal enhancement is dependent on the surface plasmons present over the nanoparticle surfaces and this enhancement is even more pronounced when two or more nanoparticles are in close proximity with each other creating the so called, hot spots. The number of these hot spots increases when the nanoparticles are present in clusters. Plasmonic hot spots can also be created on the 3D structures by deposition of thin films over them by taking advantage of the 3D hierarchy of the substrates [27].

In this work, we demonstrate the use of natural plant leaves as facile, low cost and eco-friendly SERS substrates for the sensitive detection of biomolecules, using L-cysteine as a representative molecule. Five different plant leaves possessing different 3D micro/nanostructures on their surface have been examined to investigate the effect of topography on the intensity of SERS signals. Two different fabrication techniques were used to fabricate the plant leaf based SERS substrates. In one approach, a thin layer of Au was sputter coated on to the surface of the leaves, while in the other approach Au nanoparticles were drop casted on to the surface of the leaves. A comparison between the two fabrication approaches is also presented in this work.

2 Experimental

2.1 Materials

For the synthesis of Au nanoparticles, chloroauric acid ($HAuCl_4$), sodium citrate and hydroquinone were purchased from Sigma Aldrich. The fixation of fresh leaves was done using glutaraldehyde, potassium dihydrogen orthophosphate, di-sodium hydrogen orthophosphate and glycerol, all of which were supplied by Sigma Aldrich. The silicon wafer (0.5 mm, N type) for the control experiments was also supplied by Sigma Aldrich. L-cysteine for the SERS studies was purchased from Merck. All the chemicals were used as received. The stock solutions used in the studies were prepared in ultrapure water (18.2 MΩ-cm) obtained from double stage water purifier (ELGA PURELAB Option-R7). Fresh leaves from the 5 different plants, Gladiolus (*Gladiolus dalenii*), Indian cress (*Tropaeolum majus*), Colocasia (*Colocasia esculenta),* water lily (*Nymphaea nauchali*) and banana (*Musa acuminata*) were collected from Indian institute of Technology Mandi, Kamand campus, Himachal Pradesh, India – (31°42'25"N 76°55'54"E).

2.2 Synthesis of Au nanoparticles

The synthesis of Au nanoparticles was done using a seed mediated procedure according to the literature report [28]. Seed solutions were first prepared by sodium citrate reduction. In this procedure, 25 mL of ultrapure H_2O and 250 μL of 1 wt% $HAuCl_4$ was added to a 250 mL flask and stirred rapidly while boiling. Then 1 mL of 1 wt% sodium citrate was added to it to initiate reduction. The growth solution was prepared by adding 100 μL of 1 wt% $HAuCl_4$ solution to 9.8 mL of ultrapure H_2O. Then 100 μL of seed solution was added to this solution and stirred rapidly at room temperature. In the next step, 20 μL of a 1 wt% sodium citrate solution was added, followed by 100 μL of 0.03 M hydroquinone. The nanoparticle formation was immediately evidenced by the colour change of the solution from yellow to rusty red.

2.3 Fabrication of SERS substrates

The important step in the fabrication of the SERS substrate is the fixation of the plant leaves as fresh leaves dehydrate quickly which leads to the change in their surface topography. The fixation of the fresh leaves was done using glutaraldehyde and any water content was replaced by glycerol using literature reports [29].

Two methods were utilized to fabricate SERS substrates. In first method, a thin layer of Au was sputter coated over the fixated leaf substrates and control Si wafer substrate using Quorum Q150R ES combined sputter coating instrument. The second method used to fabricate the SERS substrates was the drop cast method, where a 100 μL suspension of Au nanoparticle solution was drop casted on to the fixated leaf substrates and on to the control substrate, and allowed to dry under ambient conditions inside a desiccator. For the SERS measurements, 1 mM solution of L-cysteine in deionized water was drop casted on to each of the substrates and allowed to dry under ambient conditions again inside a desiccator.

2.4 Characterisation

High resolution transmission electron microscopy (HRTEM) images were obtained using FEI Tecnai G2 200 S-Twin microscope operating at 200 kV. Field emission scanning electron microscopy (FESEM) was done using JFEI Nova Nano SEM-450 microscope. Raman spectroscopic measurements for the SERS studies were performed using Horiba LabRAM high resolution UV-VIS-NIR Raman spectrometer instrument equipped with a confocal microscope. The samples were analyzed using a 633 nm HeNe laser employing a 20X objective lens to achieve an average spot size of 3 μm diameter. The integration time for the analysis was 10 s and the samples were scanned for 5 times under 10% of total laser power (17 mW).

3 Results and discussion

In this work, we report a strategy for the sensitive detection of biomolecules by SERS technique, employing natural plant leaves deposited with Au as substrates. As a model biomolecule, the amino acid L-cysteine, whose structure is shown in Figure 1, has been used for our investigations. In our studies, we used five different plant leaves, including gladiolus (*Gladiolus dalenii*), Indian cress (*Tropaeolum majus*), colocasia (*Colocasia esculenta*), water lily (*Nymphaea nauchali*) and banana (*Musa acuminata*), that possess different surface topographical features ranging from hierarchical 3D micro/nanostructures to a near-planar surface devoid of intricate surface textures. Prior to the experiments, in order to prevent the leaves from dehydration, which results in the curling and loss in the surface activity, the leaves were fixated with the glutaraldehyde and water content was replaced by the glycerol to make them stable and preserve the surface topography. In this way, the leaves can be made stable for a long duration of time, which makes it usable for practical applications, and no apparent damage happened to the leaves during our observation for a period of over 3 months. Furthermore, the fixation process also helps in directly observing the surface topography of the leaves using microscopic techniques without any damage to the leaves or instrument. The photograph of the plant leaves in their natural habitat along with the SEM images of the fixated leaves is shown in Figure 2. It can be evidenced from the SEM images that the surface topography of each of the plant leaf is different from the other, wherein *G. dalenii*, *T. majus* and *C. esculenta* show hierarchical micro/nanostructures on the surface, while *N. nauchali* and *M. acuminata* have a relatively planar structure with minimal surface features. In particular, the surface topography of *G. dalenii* and *C. esculenta* are somewhat similar (as can be seen from the respective magnified SEM images shown as inserts) possessing intricate hierarchical 3D micro/nanostructures. In addition, they contain micropapillae arrays of nanofolds distributed throughout the surface of the leaves. The surface topography of *T. majus* shows the presence of several short wax tubules leading to a rich surface texture. On the other hand, *N. nauchali* and *M. acuminata* show a relatively planar surface topography with micrometer level ridges and panes running adjacent to each other.

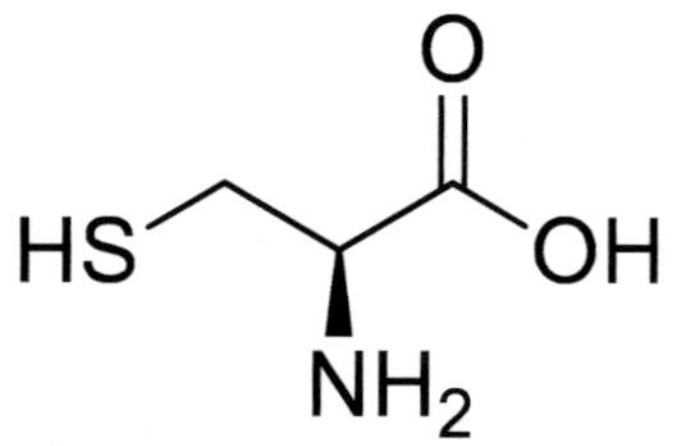

Figure 1. Chemical structure of L-cysteine.

In order to use the fixated plant leaves as substrates for SERS, we had to deposit a plasmonic material on to them. In our work, we used Au for this purpose, as it has been widely reported as a very good material for SERS applications. Specifically, we performed the deposition of Au on to the fixated plant leaves by two different methods, sputter coating and drop-cast deposition. In sputter coating method, a thin film of Au was deposited on to the fixated leaves, wherein the intrinsic surface topography of the leaves could directly be used to generate electromagnetic hot spots for SERS activity. On the other hand, in drop-cast deposition method, we used separately synthesized Au nanoparticles and deposited them on to the fixated leaf substrates, wherein the nanoparticles gets entrapped within the surface micro/nanostructures in clusters and can serve as electromagnetic hot spots for SERS activity. The TEM image along with the EDAX spectrum of the as-synthesized Au nanoparticles used in our work is shown in Figure 3, while Figure 4 shows the SEM images of the Au deposited fixated plant leaves.

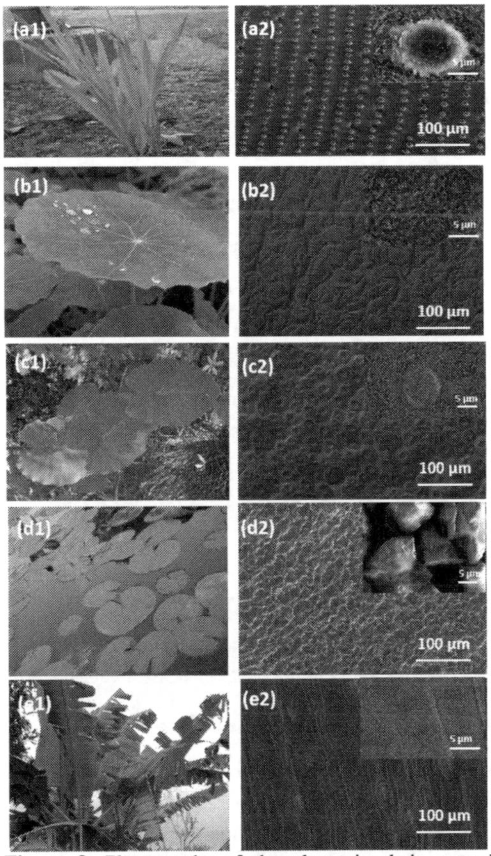

Figure 2. Photographs of the plants in their natural habitat (1) and the corresponding SEM images of the fixated leaves (2) of (a) *G. dalenii*, (b) *T. majus*, (c) *C. esculenta*, (d) *N. nauchali* and (e) *M. acuminata*. Insert shows the magnified SEM image of the fixated leaves.

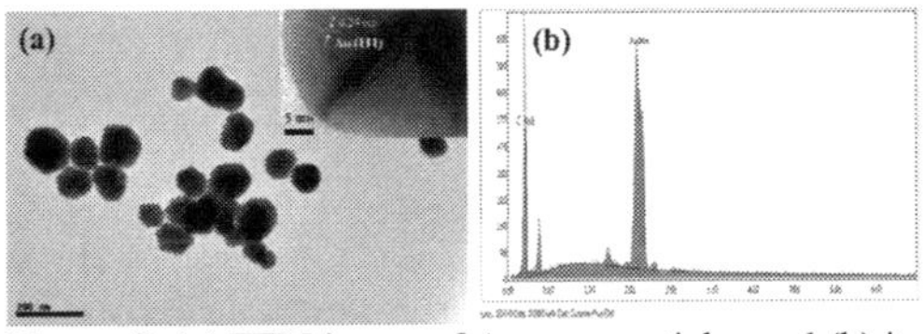

Figure 3. (a) TEM image of Au nanoparticles and (b) its corresponding EDAX spectra.

The above mentioned fixated plant leaves, after Au deposition, were used as SERS substrates for the sensitive detection of the biomolecular analyte, L-cysteine. The analyte was drop-casted on to each of the substrates in the same volume and concentration, so that the obtained SERS results could be directly compared and correlated with the surface topographical features of the plant leaves. In addition, a smooth Si wafer, deposited with Au, was used as a control substrate. The SERS spectra of L-cysteine obtained for the sputter coated and drop casted plant leaf based substrates are shown in Figure 5. It can be observed that all the peaks obtained in the spectra correspond to the signature peaks of L-cysteine as reported in literature [30] and no additional peaks, possibly arising from the leaf substrate or background, were evidenced. All the spectra showed a typical peak at 500 cm^{-1}, attributable to the S-S stretching vibration in the L-cysteine molecule. Also the characteristic bands of L-cysteine at 619 cm^{-1} (C-S stretching), 681 cm^{-1} (C-S stretching), 1201 cm^{-1} (C_α-H bending) and 1201 cm^{-1} (CH_2 twisting) could be evidenced in the spectra . Slight blue shifting of the characteristic peaks was observed, which can be due to the charge transfer interaction between L-cysteine and the Au nanoparticles.

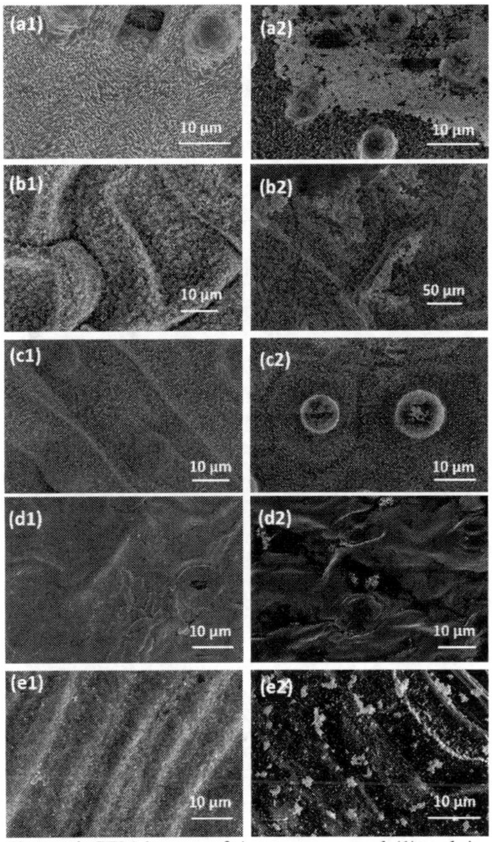

Figure 4. SEM images of Au sputter coated (1) and Au nanoparticles drop casted (2) on fixated plant leaves ((a) *G. dalenii*, (b) *T. majus*, (c) *C. esculenta*, (d) *N. nauchali* and (e) *M. acuminata*).

We examined the SERS performance of the Au deposited leaf surfaces of *G. dalenii, T. majus, C. esculenta, N. nauchali* and *M. acuminata*. All the chosen leaf substrates have different hierarchical structures naturally, which can be related to the enhanced SERS signals from the fabricated substrates. Figure 5 (a) shows the SERS spectra of L-cysteine on sputter coated leaf substrates and Figure 5 (b) shows the variation of SERS intensity of the peak at 500 cm^{-1} for each of the plant leaves. As seen clearly in the spectra, all the leaf substrates show significant enhancement in the signals as compared to the smooth Si wafer (control substrate). In addition, there is a significant difference between the performances of the SERS substrates based on different leaves. The SERS performance based upon the intensity of the signals obtained from the different substrates followed order: *G. dalenii* > *M. acuminata* > *N. nauchali* > *C. esculenta* > *T. majus*. Distinguished based on the contrast difference between the SEM images (Figure 4) it can be seen that the Au layer covered the whole surface including 3D hierarchical structures, in all the cases. Among all the fabricated substrates, *G. dalenii* leaves provide the best results wherein the papilla apex is present in ogive-like structure containing some sharp edges and sub-micron wax tubules. The other leaves with intricate hierarchical micro/nanostructures, *C. esculenta* and *T. majus* containing the sub-micrometer wax tubules but without the well-defined ogive-like structures, also show good enhancement but lesser than the *G. dalenii*. For the enhancement in the signals in these cases, we propose that the presence of ogive-like structures with sharp edges and the sub-micron tubular structures are responsible. These structures causes increase in the electric field intensity and subsequently the enhanced SERS performance. Notably in the *M. acuminata*, where there are well defined ridges and panes, arranged almost in parallel, with an average width of 5 microns extend from the bottom to the top, are responsible for the increase of electric field intensity and the formation of electromagnetic hot spots, contributing to SERS enhancement. Similarly in the case of *N. nauchali*, the micrometer level folds and the micrometer level 3D folding is responsible for the enhancement.

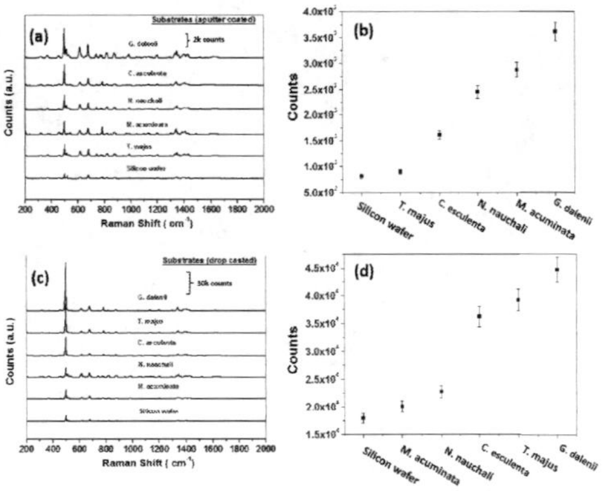

Figure 5. SERS spectra of L-cysteine for the different leaf substrates: (a) sputter coated and (c) drop casted; (b) and (d) show the variation of SERS intensity of the peak at 500 cm^{-1}, for sputter coated and drop casted plant leaf based substrates, respectively shown along with the smooth Si wafer, used as control.

Figure 5 (c) shows the SERS spectra of all the fabricated leaf substrates along with the control Si substrate drop casted with the Au nanoparticles and Figure 5 (d) shows the variation of SERS intensity of the peak at 500 cm^{-1} for each of the plant leaves. Strong enhancement in the Raman signal is clearly evident due to the presence of the Au nanoparticle clusters on the fixated plant leaf based substrates. As observed from the spectra acquired from the drop casted SERS substrates, the leaves possessing intricate hierarchical 3D micro/nanostructures on their surface (*G. dalenii, T. majus,* and *C. esculenta,*) show better enhancement of the signals as compared to those with relatively planar surfaces (*N. nauchali* and *M. acuminata*). The intricate micro/nanostructures found on *G. dalenii, T. majus,* and *C. esculenta* lead to the entrapment of larger number of nanoparticles to form nanoclusters and thereby enhance the electric field intensity of the electromagnetic hot spots, which is responsible for increase in the Raman signal intensities of the analyte molecule. In this case as well, the Au nanoparticle decorated *G. dalenii* leaves provided the best enhancements in SERS signals, followed by the *C. esculenta, T. majus, N. nauchali* and *M. acuminata*. Based on our results, we can conclude that the three dimensional distribution of the Au nanoparticle clusters formed on the ogive-like protrusions on the surface of *G. dalenii* was better than the 3D arrangement of Au nanoparticle clusters on the micro/nanostructures present on the surfaces of *T. majus and C. esculenta*. The relatively low SERS enhancement on *N. nauchali* and *M. acuminata* could be attributed to their near-planar surface texture, which leads to the dispersion of Au nanoparticles and subsequently the formed nanoclusters are small comprising of fewer numbers of nanoparticles. In both sputter coated and drop casted cases, the smooth Si wafer, which is devoid of any micro/nanostructures on its surface had the least SERS performance in the detection of L-cysteine.

The enhancement factor (EF) for the plant leaf based SERS substrates was calculated by the following equation [31]:

$$EF = \frac{I_{SERS}/C_{SERS}}{I_{RS}/C_{RS}} \tag{1}$$

where, I_{SERS} is the intensity of enhanced spectrum, C_{SERS} is the concentration of analyte in enhanced spectrum, I_{RS} is the intensity of the Raman spectra of L-cysteine alone and C_{RS} is the concentration of L-cysteine without any substrate. The EF values determined for the different substrates are shown in Figure 6.

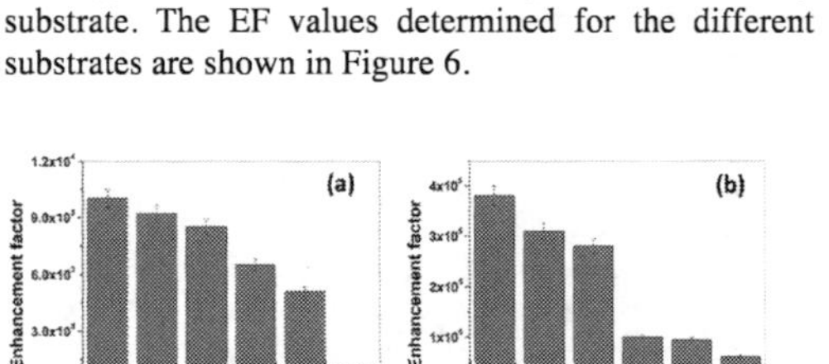

Figure 6. Enhancement factors for (a) sputter coated and (b) drop casted substrates plant leaf based SERS substrates along with the smooth Si wafer used a control.

The enhancement factors reported in this work correlate well with the corresponding values reported in literature for similar systems such as butterfly wings, plant leaves, flower petals, and other biological surfaces used for SERS applications [23, 25, 32]. In addition, the Au deposited fixated leaf substrates had reproducible SERS signals.

4 Conclusions

In this work, we have demonstrated the sensitive detection of L-cysteine (a model biomolecule) using SERS technique by employing natural plant leaves as facile, low cost and eco-friendly substrates. In particular, we have reported on the influence of the surface topography of five different plant leaf based substrates (deposited with Au) on the SERS performance. Furthermore, we have also compared the effect of sputter coated Au thin films with drop casted Au nanoparticles on the leaf substrates. Based on our results, we claim that the leaves possessing intricate hierarchical 3D micro/nanostructures show higher SERS enhancement compared to the leaves having a near-planar surface. In addition, leaves having drop-casted Au nanoparticle clusters performed better than the leaves sputter coated with a thin Au film.

Acknowledgements

We are thankful to Advanced Materials Research Center (AMRC), IIT Mandi for laboratory and the characterization facilities. VK acknowledges the financial support from DST, India under INSPIRE Faculty Award. VS acknowledges scholarship from MHRD, India.

References

1 M. Karplus and J. A. McCammon, Nature Structural & Molecular Biology **9** (9), 646 (2002).
2 P. Kask, K. Palo, D. Ullmann, and K. Gall, Proceedings of the National Academy of Sciences **96** (24), 13756 (1999).
3 P. Stepnicka, *Ferrocenes: ligands, materials and biomolecules*. (John Wiley & Sons, 2008).
4 C. A. Heid, J. Stevens, K. J. Livak, and P. M. Williams, Genome research **6** (10), 986 (1996); P. V. Hornbeck, Current protocols in immunology, 2.1. 1 (1991).
5 C. Herrmann and M. Reiher, in *Atomistic Approaches in Modern Biology* (Springer, 2006), pp. 85; S. Song, Y. Qin, Y. He, Q. Huang, C. Fan, and H.-Y. Chen, Chemical Society Reviews **39** (11), 4234 (2010).
6 G. E. Schulz and R. H. Schirmer, *Principles of protein structure*. (Springer Science & Business Media, 2013).
7 I. Nabiev, G. Chumanov, and R. Efremov, Journal of Raman spectroscopy **21** (1), 49 (1990).
8 D. A. Long, New York, 1 (1977).
9 N. Colthup, *Introduction to infrared and Raman spectroscopy*. (Elsevier, 2012).
10 B. Sharma, R. R. Frontiera, A.-I. Henry, E. Ringe, and R. P. Van Duyne, Materials today **15** (1), 16 (2012).
11 B. Sharma, M. F. Cardinal, S. L. Kleinman, N. G. Greeneltch, R. R. Frontiera, M. G. Blaber, G. C. Schatz, and R. P. Van Duyne, MRS bulletin **38** (08), 615 (2013).
12 M. Fleischmann, P. J. Hendra, and A. McQuillan, Chemical Physics Letters **26** (2), 163 (1974); D. L. Jeanmaire and R. P. Van Duyne, Journal of Electroanalytical Chemistry and Interfacial Electrochemistry **84** (1), 1 (1977).
13 A. Campion and P. Kambhampati, Chemical Society Reviews **27** (4), 241 (1998).
14 H. Ko and V. V. Tsukruk, small **4** (11), 1980 (2008); L. Zhang, X. Gong, Y. Bao, Y. Zhao, M. Xi, C. Jiang, and H. Fong, Langmuir **28** (40), 14433 (2012).
15 P. L. Stiles, J. A. Dieringer, N. C. Shah, and R. P. Van Duyne, Annu. Rev. Anal. Chem. **1**, 601 (2008).
16 D. Cialla, A. März, R. Böhme, F. Theil, K. Weber, M. Schmitt, and J. Popp, Analytical and bioanalytical chemistry **403** (1), 27 (2012).
17 A. Tao, F. Kim, C. Hess, J. Goldberger, R. He, Y. Sun, Y. Xia, and P. Yang, Nano letters **3** (9), 1229 (2003).
18 R. A. Tripp, R. A. Dluhy, and Y. Zhao, Nano Today **3** (3), 31 (2008).
19 G. L. Liu and L. P. Lee, Applied Physics Letters **87** (7), 074101 (2005).
20 P. Rajapandiyan and J. Yang, Journal of Raman Spectroscopy **45** (7), 574 (2014).
21 C. H. Lee, M. E. Hankus, L. Tian, P. M. Pellegrino, and S. Singamaneni, Analytical chemistry **83** (23), 8953 (2011).
22 V. Sharma, S. Kumar, K. L. Reddy, A. Bahuguna, and V. Krishnan, Journal of Molecular and Engineering Materials **4** (01), 1640006 (2016); V. Sharma, M. Sharma, S. Kumar, and V. Krishnan, Flora-Morphology, Distribution, Functional Ecology of Plants **224**, 59 (2016).
23 N. L. Garrett, P. Vukusic, F. Ogrin, E. Sirotkin, C. P. Winlove, and J. Moger, Journal of biophotonics **2** (3), 157 (2009).
24 J.-A. Huang, Y.-L. Zhang, Y. Zhao, X.-L. Zhang, M.-L. Sun, and W. Zhang, Nanoscale **8** (22), 11487 (2016); S.-Y. Chou, C.-C. Yu, Y.-T. Yen, K.-T. Lin, H.-L. Chen, and W.-F. Su, Analytical chemistry **87** (12), 6017 (2015).
25 V. Sharma, S. Kumar, A. Jaiswal, and V. Krishnan, ChemistrySelect (2017), DOI: 10.1002/slct.201601451

[26] H. Ko, S. Singamaneni, and V. V. Tsukruk, Small **4** (10), 1576 (2008).
[27] M. Chen, I. Y. Phang, M. R. Lee, J. K. W. Yang, and X. Y. Ling, Langmuir **29** (23), 7061 (2013).
[28] S. D. Perrault and W. C. Chan, Journal of the American Chemical Society **131** (47), 17042 (2009).
[29] H. Ensikat, P. Ditsche-Kuru, W. Barthlott, and A. Méndez-Vilas, Microscopy: Science, technology, applications and education **1** (13), 248 (2010); H. Ensikat and W. Barthlott, Journal of microscopy **172** (3), 195 (1993).
[30] H. Lee, M. S. Kim, and S. W. Suh, Journal of Raman spectroscopy **22** (2), 91 (1991).
[31] V. Sharma, N. Sinha, S. Dutt, M. Chawla, and P. F. Siril, Journal of colloid and interface science **463**, 180 (2016).
[32] Z. Mu, X. Zhao, Z. Xie, Y. Zhao, Q. Zhong, L. Bo, and Z. Gu, Journal of Materials Chemistry B **1** (11), 1607 (2013); X. Liu, C. Zong, K. Ai, W. He, and L. Lu, ACS applied materials & interfaces **4** (12), 6599 (2012).

Determination of the Glass Transition Temperature of Freestanding and Supported Azo-Polymer Thin Films by Thermal Assisted Atomic Force Microscopy

Elena Chernykh[1], *Sergey* Kharintsev[1,2,*], *Alexandr* Fishman[1], *Alexander* Alekseev[3,4], *Myakzuym* Salakhov[1,2]

[1]Department of Optics and Nanophotonics, Institute of Physics, Kazan Federal University, Kremlevskaya, 16, Kazan, 420008, Russia
[2]Institute of Perspective Technologies, Tatarstan Academy of Sciences, Baumana, 20, Kazan, 420111, Russia
[3]National Laboratory Astana, Nazarbayev University, Kabanbay batyr ave., 53, Astana, 010000, Kazakhstan
[4]CSF ECB&MST, National Research University MIET, 124498 Moscow, Russia

Abstract. In this paper we introduce and apply the method for determination of the glass transition temperature of the sub-100 nm thick freestanding and supported polymer films based on thermally assisted atomic force microscopy (AFM). In proposed approach changes of the phase of an oscillating AFM cantilever are used to determine glass transition temperature. An anomalous decrease of the glass transition temperature for both free-standing and supported azobenzene-functionalized polymer thin films is shown.

1 Introduction

Azobenzene-functionalized polymers are one of the most promising materials in polymer photonics and organic electronics. Due to *trans*-to-*cis* photoisomerization, azo-chromophores found applications in formation of surface relief gratings and study of nonlinear optical effects.[1-4] These effects are widely used in practice, for example for optical switching and storage,[5-6] light filtering[7-10] and frequency conversion.[11] In particular, bottlenecks of optical near-field memory based on azo-polymers are cycling endurance, data retention, and thermal stability. Photoinduced alignment of the azo-dyes in glassy polymers is affected by the thickness of a polymer thin film, since the glass transition temperature T_g critically drops when the film thickness decreases.[12] In addition, T_g is influenced by the interface interaction and, therefore, type of a substrate plays role as well.[13,14]

The integrity and thermal stability of the oriented state of polymer thin films are greatly sensitive to the glass transition temperature, at which an intrinsic structure of the polymer changes from a hard disordered solid to a rubber-like bulk. In turn, this influences on the increased mobility of a surface layer that underlies the most observable photo-induced effects in the azo-polymers. The determination of the glass transition temperature has been the subject of through studies over last 50 years. Nowadays, there are many methods for determining the glass transition temperature of bulk amorphous and liquid crystalline polymers such as differential scanning calorimetry[15], dynamic mechanical analysis[16], Raman scattering[17], Brillouin scattering[18] and others. However, most of these methods have a low sensitivity to the sub-100 nm thick polymer films. Atomic force microscopy is a promising technique, which provides information about glass transitions through a mechanical behavior of the AFM cantilever. In this paper, we determine the glass transition temperature of freestanding and supported azo-polymer films using thermal assisted atomic force microscopy.

2 Experimental section

Herein, we study thin films of epoxy-based oligomers containing hydroxyl groups with covalently attached nitroazobenzene chromophores, further it is referred to as CFAO. A chemical structure of this molecule is given in Fig. 1(a). The solution was prepared by dissolving 5 wt. % CFAO compound in 1 ml cyclohexanone. Polymer films with different thicknesses in the range of 5-500 nm were fabricated. The thin films were produced by spin coating 1 μl CFAO solution onto a glass substrate at a rotation rate of 10 000 rpm for 2 min. Freestanding films were prepared by centrifugation on a KBr substrate. Then substrate was dissolved in distilled water and after that thin polymer films were picked up with a mesh copper grids as shown in Fig. 1(b). The film thickness was determined by atomic force microscopy in contact mode. In order to minimize errors in determining the film thickness, we have built the dependence of the depth of lithographically produced "grooves" on applied force of the AFM cantilever. The film thickness was defined as a value that meets a plateau on this curve (see Fig.1(c)).

* Corresponding author: skharint@gmail.com

DOI: 10.1051/epjconf/201713900007

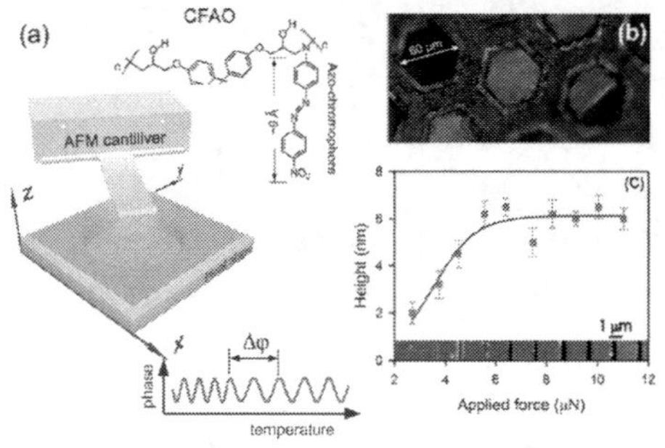

Fig. 1. a) experimental setup and chemical formula of CFAO; b) optical image of the freestanding CFAO film; c) plot of the groove depth on applied contact force.

The phase measurements were performed with an atomic force microscope Ntegra Prima (NT-MDT) in tapping mode. We used silicon AFM cantilever with the curvature radius of 10 nm, the resonant frequency of 150 kHz and the force constant of 5 N/m (NSG01, NT-MDT). Polymer films were heated with the thermo stage (SU045NTF, NT-MDT) with an accuracy of 0.05 °C within the range from room temperature to 150 °C. A schematic drawing of the experiment is given in Fig. 1(a). In our approach, the atomic force microscope was used as a dynamic mechanical analyzer of the AFM cantilever phase shift during *in-situ* heating/cooling at a rate of 5 °C/min (Fig. 1(a)).

3 Results and discussion

Due to the fact that the phase of the cantilever oscillations is sensitive to the surface properties, we can use this parameter to monitor the glass transition of polymer films. It is known that cantilever phase shift φ is determined by energy dissipated during tip-sample interaction: $\sin \varphi \sim E_{dis}$, where E_{dis} is energy dissipated during one period of oscillation.[19]

The phase of the cantilever oscillations is influenced by viscoelasticity, friction, adhesion etc. which leads to energy dissipation. This can be attributed to the loss of energy of the cantilever, which is associated with a change in the Gibbs free energy of the polymer: $\Delta G = \Delta H - T\Delta S$. At the glass transition temperature, changes in enthalpy ΔH and entropy ΔS of the CFAO thin films are changed and it can lead to reversible jumps of the AFM phase during heating (red curve) and cooling (blue curve).[20] For stable and reliable AFM measurements with *in-situ* heating/cooling, we heat or cool the polymer step-by-step by a value of 20 °C at a linear thermal rate of 5 °C/min. This stepwise regime permits the sample to be in a near-equilibrium state. Using this approach, T_g can be defined as the temperature at which the jump on phase curve occurs during heating/cooling, as shown in Fig. 2(a).

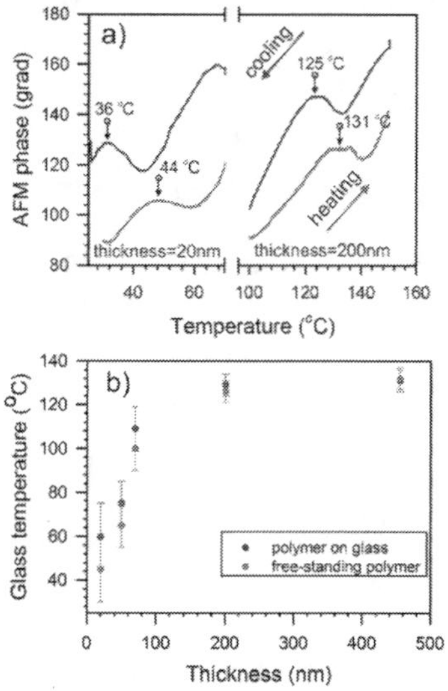

Fig. 2. A plot of the AFM phase vs. temperature of the CFAO thin film with thickness of 20 nm and 200 nm (a), a dependence of the glass transition temperature on the thickness of freestanding and supported CFAO films.

This figure shows results of measurements of the glass transition temperature for both 20 nm and 200 nm thick CFAO films. The T_g value for the bulk film coincides with the results obtained with other methods and is equal to ~130 °C. For a thiner CFAO film we observe the decrease of T_g down to roughly 40 °C. The depression of the glass transition temperature with the decrease of film thickness was shown in numerous studies.[13,17,21] Thus, our results are in agreement with earlier published ones. Fig. 2(b) demonstrates the influence of thickness on T_g for both free-standing and supported CFAO thin films. As follows from this figure, the T_g behaviours for both freestanding and supported CFAO thin films are close to each other. It means that a role of the substrate in our case is negligable. This was experimentally confirmed by a poor wettability of the CFAO film. Indeed, very thin films with the thcikness of less than 7 nm exhibit a ragged stucture on the subsrate surface.

4 Conclusions

The glass transition temperature of CFAO thin films can be determined by using thermal assisted atomic force microscopy in tapping mode. The applied method is based on high sensitivity of the oscillating cantilever phase to changes of the surface properties. This method gives reliable and robust estimations for the glass transition temperature for very thin films with the thickness of less than 20 nm.

Acknowledgment

This work was financially supported by the Russian Foundation for Basic Research (No. 15-42-02339). This work was done using equipment of Federal Center of Shared Equipment of Kazan Federal University.

References

1. N.A. Nikonorova, M.Yu. Balakina, O.D. Fominykh, M.S. Pudovkin, T.A. Vakhonina, R. Diaz-Calleja, A.V. Yakimansky, Chem. Phys. Lett. **522**, 114 (2012)
2. Photo-reactive organic thin films; Z. Sekkat, W. Knoll. Eds; Academic press: Elsevier Science, USA, 2002.
3. Z. Sekkat, M. Buchel, H. Orendi, H. Menzel, W. Knoll, Chem. Phys. Lett. **220**, 497 (1994).
4. A. Natansohn, P. Rochon, Chem. Rev. **102**, 4139 (2002)
5. A. Natansohn, P. Rochon, M. Pezolet, P. Audet, D. Brown, S., Macromolec. **27**, 2580 (1994)
6. W.Shi, Yu. J. Ding, X. Mu, Appl. Phys. Lett. **79**, 3749 (2001)
7. Y. Wang, O.Y.-H. Tai, C.H. Wang, J. Chem. Phys. 2005. **123**, 704 (2005)
8. M.S. Ho, A. Natansohn, Macromolec. **28**, 6124 (1995)
9. S.S. Kharintsev, A.I. Fishman, S.G. Kazarian, I.R. Gabitov, M.Kh Salakhov, ACS Photonics. **1**, 1025 (2014)
10. S.S. Kharintsev, A.I. Fishman, S.K. Saikin, S.G. Kazarian, Nanoscale **8**, 19867 (2016).
11. C. Fiorini, F. Charra, J.-M. Nunzi, P. Raimond, J. Opt. Soc. Am. B. **14**, 1984 (1997)
12. J.L. Keddie, R.A.L. Jones, R.A. Cory, Faraday Discuss. **98**, 219 (1994)
13. D.S. Fryer, P.F. Nealey, J.J. de Pablo, Macromolec. **33**, 6439 (2000)
14. J.A Torres, P.F. Nealey, J.J. de Pablo, Phys. Rev. Lett. **85**, 3221 (2000)
15. E. Dargent, C. Cabot, J.M. Saiter, J. Bayard, J. Grenet, Jour. of Therm. Analys. **47**, 887 (1996)
16. B. Twombly, D.D. Shepard, Instrum. sci. and technolog. **22**, 259 (1994)
17. H. Liem, J. Cabanillas-Gonzalez, P. Etchegoin, D.D.C Bradley, Journ. of Phys.: Cond. Matt. **16**, 721(2004)
18. J.A. Forrest, K. Dalnoki-Veress, J.R. Dutcher, A.C. Rowat, J.R. Stevens, Disordered materials and interfaces. **407**, 131 (1996)
19. J. P. Cleveland, B. Anczykowski, A.E. Schmid, V.B. Elings, Appl. Phys. Lett. **72**, 2613–2615 (1998).
20. M. Meincken, L.J. Balk, R.D. Sanderson, Surf. Interface Anal. **35**, 1034 (2003)
21. D. Liu, R.O. Orozco, T. Wang, Phys. Rev. E. **88**, 22601 (2013)

EPJ Web of Conferences **139**, 00008 (2017) DOI: 10.1051/epjconf/201713900008
NANOP 2016

How Plasmonic excitation influences the LIPSS formation on diamond during multipulse femtosecond laser irradiation ?

Ahmed Abdelmalek[1], *Zeyneb Bedrane*[1,*], *El-Hachemi Amara*[2], *Shane M. Eaton*[3] *and Roberta Ramponi*[3]

[1]Theoretical Physics Laboratory, Physics Dpt. Sciences Faculty,Tlemcen University, B.P. 119, 13000 Tlemcen, Algeria
[2]Laser Material Processing Team, Centre for Advanced Technologies Development, CDTA, PO Box 17 baba-Hassan, 16303 Algiers, Algeria
[3]Physics Dpt., Politecnico di Milano, Italy

Abstract. A generalized plasmonic model is proposed to calculate the nanostructure period induced by multipulse laser femtosecond on diamond at 800 nm wavelengths. We follow the evolution of LIPSS formation by changing diamond optical parameters in function of electron plasma excitation during laser irradiation. Our calculations shows that the ordered nanostructures can be observed only in the range of surface plasmon polariton excitation.

1 INTRODUCTION

Laser Induced Periodic Surface Structure, also called "ripples or nanograting" are part of morphologies that have focused the interest of several researchers and have been on the heart of a lot of studies since the first experiment done by Birnbaum in 1965 where two types of training LIPSS formation were observed during the irradiation of dielectric, semiconductor and metals [1, 2, 3] by multipulse femtosecond laser linearly polarized, called low spatial frequency LIPSS (LSFL) and high spatial frequency LIPSS (HSFL). LSFL formation for the majority of materials is well explained by the interference between the incident laser wave and the plasmon polaritons SPP surface [4, 3], which involves a periodic spatial modulation of the energy deposited in the irradiated surface [5, 6], this leads to the formation of nanostructures with periods lower or near to λ_L.

But HSFL formation remain a misunderstood phenomenon and it is still under discussion. There are several models proposed and many studies were done to identify the origin of formation of this such type of LIPSS on dielectric and semiconductor, induced by multipulse laser femtosecond at low fluency [7, 3, 8, 9, 10]. In such cases, the nonthermal interaction and plasmon excitation play a very important role, where the HSFL's period is very lower to the incident laser wavelength. Masataka Shinoda *et al* [10] noted a period of 146±7 nm on diamond irradiated by 120 fs, 250 kHz at λ_L =800 nm. Straub *et al* [11] find a period of 145 nm on the Si (100) at λ_L =800 nm.

A recent explanation of HSFL formation was proposed by Miyazaki *et al* [9] and Miyaji *et al* [8] where they proposed that by increasing the number of low fluence pulses, the periodic nanostructures HSFL can be developed through the nonthermal structure change of the treated surface. The nanoscale ablation due to strong nearfields appears around of sweeling followed by a plasmonic excitation (SPP) to improve the nano-periodicity of irradiated area.

In this work, we illustate the plasmon excitation advantage to training HSFL formation through a general plasmon model that can follow the evolution of training LIPSS formation by changing diamond optical parameters in function of electron plasma excitation.

* Corresponding author: zeyneb_bedrane@yahoo.fr

2 THEORETICAL MODEL

We propose a generalized plasmon model which allows to follow the development of nanostructures period during diamond irradiation by femtosecond laser pulse at 800 nm wavelength. We consider the air as the dielectric environment.

We use the plasmon model in the case of dielectric thanks to the specific advantage of femtosecond laser wich allow to create, during its interaction with material, a thin layer that has a metallic character (it will be noted for the following pseudo-metal layer).

The surface plasmon polariton dispersion relation SPP is [12]:

$$\omega^2 = c^2 K_{sp}^2 \frac{\left(\varepsilon_{pm} + \varepsilon_d\right)}{\varepsilon_{pm}\varepsilon_d} \quad (1)$$

Where ε_{pm} is the dielectric function of the pseudo-metal layer, ε_d is the dielectric constant of the surrounding the pseudo-metallic layer ($\varepsilon_{air} = 1$ and $\varepsilon_{diamond} = 5.7$ at 800 nm [10]).

$K_{sp} = K_{sp1} + iK_{sp2}$ Is the plasmon propagation number, $K_{sp1} = \frac{2\pi}{\lambda_{sp}} = \frac{\pi}{\Lambda}$ (λ_{sp} is the plasmon wavelengh, and Λ is the nanostructure's period induced in the diamond surface).

The dielectric function of the pseudo-metal layer is given through the Drude's model [13]:

$$\varepsilon_{pm} = \varepsilon_{pm_1} + i\varepsilon_{pm_2} = 1 + \left(\varepsilon_{diamond} - 1\right)\left(1 - \frac{n_{eh}}{n_0}\right) - \frac{\omega_p^2}{\omega^2} \times \left(\frac{1}{1 + \frac{i}{\omega\tau_{e^-e^-}}}\right) \quad (2)$$

where n_0 is the valence band electron concentration supposed equal to 10^{23} cm^{-3}. $\tau_{e^-e^-}$ is the electron-electron collision time supposed equal to 1 fs.

$\omega_p = \sqrt{\frac{n_{eh} e^2}{\varepsilon_0 m_{opt} m_e}}$ is the plasma frequency

($\varepsilon_0 = 8.85 \times 10^{-12}$ F.m and the optical effective mass $m_{opt} = 0.28$ [13, 14]).

M. Straub *et al* [11] have determined the grooves's period of Si after formalizing formulas of K_{sp_1} and K_{sp_2} via the ejection of the critical density of plasmon excitation n_{cr} for the objective to simplify the calculation such as $\varepsilon_{pm_1}\left(n_{cr}\right) = 0$ (in our case $n_{cr} = 3.210^{21}\, cm^{-3}$ at λ = 800 nm), so this formulas become special case just for the property of Si and the wavelength used.

In our case, we will try in this paper to build a general formula of K_{sp_1} and K_{sp_2} valid for all semiconductor material and dielectrics and for all femtosecond laser wavelengths.

In the one hand, according to the formula (1)

$$K_{sp}^2 = \left(\frac{\omega}{c}\right)^2 \times \left(\frac{\varepsilon_{pm}\varepsilon_d}{\varepsilon_{pm} + \varepsilon_d}\right)$$

In the other hand, we have

$$K_{sp}^2 = \left(K_{sp_1} + iK_{sp_2}\right)^2 = K_{sp_1}^2 - K_{sp_2}^2 + 2\mathrm{i}K_{sp_1}K_{sp_2}$$

$$\frac{1}{K_{sp}^2} = \frac{c^2}{\omega^2}\left(\frac{1}{\varepsilon_d} + \frac{1}{\varepsilon_{pm}}\right) = \frac{c^2}{\omega^2}\left(\frac{1}{\varepsilon_d} + \frac{1}{\left(\varepsilon_{pm_1} + i\varepsilon_{pm_2}\right)}\right)$$

$$\frac{1}{K_{sp}^2} = \frac{c^2}{\omega^2}\left(\frac{\varepsilon_{pm_1}^2 + \varepsilon_{pm_2}^2 + \varepsilon_d\varepsilon_{pm_1} - i\varepsilon_d\varepsilon_{pm_2}}{\varepsilon_d\left(\varepsilon_{pm_1}^2 + \varepsilon_{pm_2}^2\right)}\right)$$

We put:

$$\alpha = \varepsilon_{pm_1}^2 + \varepsilon_{pm_2}^2 + \varepsilon_d\varepsilon_{pm_1}$$

$$\beta = \varepsilon_d\varepsilon_{pm_2}$$

$$\gamma = \varepsilon_d\left(\varepsilon_{pm_1}^2 + \varepsilon_{pm_1}^2\right)$$

Which implies:

$$K_{sp}^2 = \frac{\omega^2}{c^2}\left(\frac{\gamma}{\alpha - i\beta}\right)$$

We have also:

$$K_{sp_1}^2 - K_{sp_2}^2 = \left(\frac{\omega^2}{c^2}\right)\left(\frac{\gamma\alpha}{\alpha^2 + \beta^2}\right) \quad (3)$$

$$2K_{sp_1}K_{sp_2} = \left(\frac{\omega^2}{c^2}\right)\left(\frac{\gamma\beta}{\alpha^2 + \beta^2}\right) \quad (4)$$

Equation (4) implies $K_{sp_2}^2 = \left(\frac{\omega^2}{2c^2 K_{sp_1}}\right)^2 \frac{\gamma^2\beta^2}{\left(\alpha^2 + \beta^2\right)^2}$

Replacing in (3), we found:

$$4K_{sp_1}^4 - 4\frac{\omega^2}{c^2}\frac{\gamma\alpha}{\left(\alpha^2 + \beta^2\right)}K_{sp_1}^2 - \left(\frac{\omega^2}{c^2}\right)^2 \frac{\gamma^2\beta^2}{\left(\alpha^2 + \beta^2\right)^2} = 0 \quad (5)$$

To find the roots of this polynomial, we put $x=K_{sp_1}^2$ and :

$$a=4\ ,\ b=-4\frac{\omega^2}{c2}\frac{\gamma\alpha}{\alpha^2+\beta^2}\ ,\ d=-\left(\frac{\omega^2}{c^2}\right)^2\frac{\gamma^2\beta^2}{\left(\alpha^2+\beta^2\right)^2}$$

then we can write (5) under the form: $ax^2+bx+d=0$

We must seek a real and positive value of K_{sp_1} which is physically accepted. We finally obtain

$$K_{sp_1}=\frac{1}{\sqrt{2}}\frac{\omega}{c}\sqrt{\frac{\gamma\alpha}{\alpha^2+\beta^2}+\frac{\gamma}{\sqrt{\alpha^2+\beta^2}}}$$

$$K_{sp_2}=\frac{\omega^2}{2c^2K_{sp_1}}\frac{\gamma\beta}{\left(\alpha^2+\beta^2\right)}$$

3 RESULTS AND DISCUSSION:

When the diamond surface is excited by multipulse laser femtosecon, a fine pseudo-metal layer is build [19] due to the non-thermal fusion (disorder) which takes place at sub-vibrational duration time [15]. It is a special property of femtosecond laser because the pulse duration is less than the electron-phonon relaxation time [16, 17, 18]. A change of states appears clearly in the optical properties of diamond such reflectivity $R=\frac{(n-1)^2+\kappa^2}{(n+1)^2+\kappa^2}$ (refraction index $n=\frac{1}{\sqrt{2}}\sqrt{\varepsilon_{pm_1}+\sqrt{\varepsilon_{pm_1}^2+\varepsilon_{pm_2}^2}}$ and the absorption index $\kappa=\frac{\varepsilon_{pm_2}}{2n}$ [20]).

Fig.1 shows the evolution of the diamond reflectivity in function of density ratio at $\lambda=800\,nm$. We notice that the reflectivity increase when the plasma density increase n_{eh} .

So, after a multipulse femtosecond laser, the diamond take the characteristic of a metal with high free electron density.

When the pulse number increase, the excited electron density n_{eh} increase too until the reflectivity reached 33% (fig. 1) which correspond to $\varepsilon_{pm}=-\varepsilon_{air}$, this is the point where the first order nanostructure can be observed with a period $\Lambda_{min}=397nm$ at $\lambda=800nm$, we will call it minimum plasmon resonance RP_{min} . This RP_{min} is caused when the plasmon tends to a coherent coupling with the incident laser at air/pseudo-metallic interface as if the original material (diamond) does not exist.

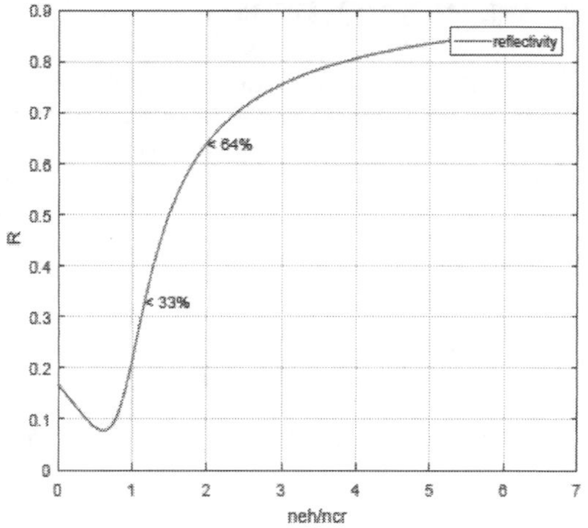

Fig. 1. evolution of the diamond reflectivity in function of density ratio at $\lambda=800nm$. 33% and 64% correspond to RP_{min} and RP_{max} respectively.

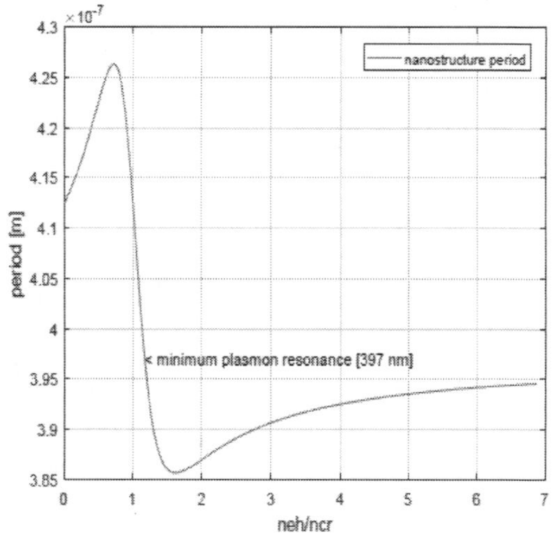

Fig. 2. behaviour of the nanostructure's period Λ in function of the density ratio at the interface air/pseudo-metallic at λ=800 nm

When the number of pulse increase again, the pseudo-metallic dielectric function decreased until reach $\varepsilon_{pm}=-\varepsilon_{dimond}$. At this point we find the smallest nanostructure period, also called as the saturation period [2], $\Lambda_{max}=160nm$ at λ = 800 nm (fig. 3), due to the maximum plasmon resonance RP_{max} at the interface pseudo-metal/diamond.

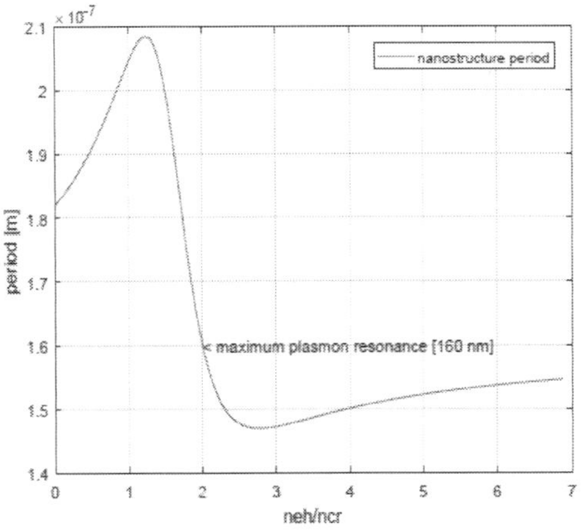

Fig. 3. behaviour of the nanostructure's period Λ in function of the density ratio at the interface pseudo-metallic/diamond at λ=800 nm.

We can conclude that there is like a transverse migration of the surface plasmon polariton SPP form the air/pseudo-metal interface to pseudo-metal/diamond during the increase of femtosecond laser pulse number and the nanostructure period decrease form Λ_{min} to Λ_{max} . So we deduce that the ordered nanostructures can be observed only in the range $[RP_{min}, RP_{max}]$.

4 Conclusion:

In this paper we have derived a generalized relations of K_{sp_1} and K_{sp_2} which can be used directly to calculate the period of nanostructures and other optical parameters at any wavelength femtosecond laser incident without need to return to basic relations (1) and (2) for all the dielectric or semiconductor material.

We have used this model to calculate the periods of nanostructures that can be observed on the diamond during irradiation by multipulse femtosecond laser at 800 nm in air room. We have determined the saturation period Λ_{max} that corresponds to the maximum plasmon resonance RP_{max} . We conclude that the periodic nanostructures can be observed only in the range $[RP_{min}, RP_{max}]$.

References

[1] T. Q. Jia, H. X. Chen, M. Huang, F. L. Zhao, J. R. Qiu, R. X. Li, Z. Z. Xu, X. K. He, J. Zhang, and H. Kuroda, , *Phys. Rev. B* **72**, 125429 (2005).

[2] S. Richter, M. Heinrich, S. Döring, A. Tünnermann and S. Nolte, *Appl Phys A* **104**, 503 (2011).

[3] M. Huang, F. Zhao, Y. Cheng, N. Xu, and Z. Xu, *ACS Nano* **3**, 12 (2009).

[4] T.H.R. Crawford and H.K. Haugen, *App. Sur. Sci.* **253,** 4970 (2007).

[5] T. Chen, H. Yang, E. Zhao, J. Qian, J. Hao, J. Han, W. Tang and H. Zhu, J. of Russ. Las. Res, **35**, 6, (2014)

[6] A. Y. Vorobyev and C. Guo, *Las. Phot. Rev.* **17**, 1 (2012)

[7] S. He, J. J. Nivas, K.K. Anoop, A. Vecchione, M. Hu, R. Bruzzese and S. Amoruso, *App. Sur. Sci.* **353**, 1214 (2015)

[8] G. Miyaji and K. Miyazaki, *App. Phys. Lett.* **103**, 071910 (2013)

[9] K. Miyazaki , G. Miyaji, *Phys. Proc.* **39,** 674 (2012)

[10] M. Shinoda, R. Gattass, and E. Mazur, *J. App. Phys.* **105**, 053102 (2009)

[11] M. Straub, M. Afshar, D. Feili, H. Seidel, and K. Konig, *J. App. Phys.* **111**, 124315 (2012)

[12] Near-field optics and surface plasmon polaritions, edited by S. Kawata (Springer, Berlin, 2001), p. 19.

[13] K. Sokolowski-Tinten and D. von der Linde, *Phys. Rev. B* **61**, 4(2000)

[14] *Diamond Electronic Properties and Applications,* edited by Lawrence S. Pan and Don R. Kania, (Springer, New York, 1995), p. 251, 252

[15] G. Sciaini, M. Harb, S. G. Kruglik, T. Payer, C.T. Hebeisen, F.J. Meyer zu Heringdorf, M. Yamaguchi, M. Horn-von Hoegen, R. Ernstorfer and R. J. Dwayne Miller, *Nature* **458|5**, 07788 (2009).

[16] S. K. SUNDARAM and E. MAZUR, **Nature mat.** | **1,** 1 (2002).

[17] C. V. Shank, R. Yen, and C. Hirlimann , *Phys. Rev. Lett.* **50**, 6 (1983).

[18] T. Zier, Eeuwe S. Zijlstra, A. Kalitsov, I. Theodonis, and M.E. Garci, J. Las. Micro / Nanoengineering **2**, 2 (2007).

[20] Handbook of Optical Constants of Solids, edited by EDWARD D. PALIK, (Academic Press, USA, 1985), p. 192.

EPJ Web of Conferences **139**, 00009 (2017) DOI: 10.1051/epjconf/201713900009
NANOP 2016

Coupled-Mode Theory derivation of the formal equivalence between a three-mode waveguide and a set of three mutually coupled single-mode waveguides

Yann G. Boucher[1-2,*]

[1]FOTON, Systèmes Photoniques, UMR CNRS 6082, Université de Rennes 1, 22305 Lannion, France
[2]ENIB, Technopôle Brest-Iroise, CS 73862, 29638 Brest, France

Abstract. The formal identification between a two-mode waveguide and a system of two mutually coupled single-mode waveguides stems from the symmetries of the evolution operator. When the gap tends to zero, the super-modes of the coupled system merge continuously into the modes of the multimode waveguide. For modelling purposes, it is very tempting to extend the analogy to three-mode waveguides (and beyond). But not without some precautions…

1 Introduction

Multimode waveguides are the core element for Spatial-Division Multiplexing (SDM), a technique where one wavelength can carry several channels of information, each one allocated to a specific modal distribution [1].

From the point of view of its effective indices, a two-mode waveguide (TMWG) can be thought of as formally equivalent to a symmetric set of two coupled single-mode waveguides (SMWG) of wavevector β, with coupling constant χ. When the gap tends towards zero, the even and odd super-modes of the coupled system, of wavevectors β_{even} and β_{odd}, merge into the fundamental and first-order of the TMWG, respectively [2]. The identification proceeds as follows:

$$\begin{cases}\beta_{even} = \beta + \chi \\ \beta_{odd} = \beta - \chi\end{cases} \Rightarrow \begin{cases}\beta = \dfrac{\beta_{even} + \beta_{odd}}{2} \\ \chi = \dfrac{\beta_{even} - \beta_{odd}}{2}\end{cases} \quad (1)$$

A similar extension of Coupled-Mode Theory (CMT) to a three-mode waveguide (3MWG) – and beyond – is tempting, but far from obvious. Investigating the ternary slab system as a reference case (whose properties can be calculated with arbitrary precision), we will observe that even in the basic symmetric configuration made of three identical SMWG, the simplest CMT model must be completed by at least one degree of freedom, in order for the propagation constants of the super-modes to be correctly accounted for.

In Section 2, we use a matrix formalism to derive the wavevectors and the transverse profile of the super-modes. In Section 3, we point out the discrepancies, before proposing several ways to recover the missing degree of freedom.

2 The slab waveguide system

The ternary slab system is depicted in Fig. 1:

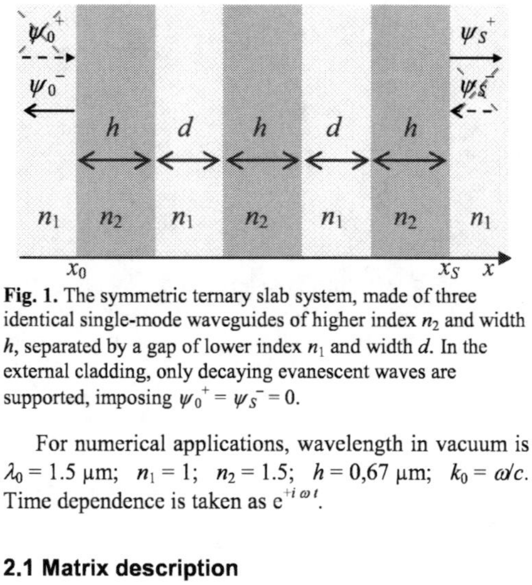

Fig. 1. The symmetric ternary slab system, made of three identical single-mode waveguides of higher index n_2 and width h, separated by a gap of lower index n_1 and width d. In the external cladding, only decaying evanescent waves are supported, imposing $\psi_0^+ = \psi_S^- = 0$.

For numerical applications, wavelength in vacuum is $\lambda_0 = 1.5$ µm; $n_1 = 1$; $n_2 = 1.5$; $h = 0{,}67$ µm; $k_0 = \omega/c$. Time dependence is taken as $e^{+i\omega t}$.

2.1 Matrix description

The matrix description of the slab system stems from the linearity of Maxwell equations [3-5]. For a given state of polarization (TE/TM), the transverse components of the wavevector of the isolated SMWG guided mode, of longitudinal propagation constant β, are either real ($\psi^+ = A\,e^{-i\kappa x}$, $\psi^- = B\,e^{+i\kappa x}$ if *propagative*) or purely imaginary ($\psi^+ = A\,e^{-\gamma x}$, $\psi^- = B\,e^{+\gamma x}$ if *evanescent*), with $\beta^2 - \gamma^2 = n_1^2 k_0^2$, $\beta^2 + \kappa^2 = n_2^2 k_0^2 > n_1^2 k_0^2$. The boundary conditions ensure that no exponential increase is allowed in the semi-infinite cladding.

* Corresponding author: yann.boucher@enib.fr

The matrix of the isolated SMWG is such as:

$$\begin{pmatrix} 0 \\ \psi_0^- \end{pmatrix} = \begin{pmatrix} m_{11} & m_{12} \\ m_{21} & m_{22} \end{pmatrix} \begin{pmatrix} \psi_S^+ \\ 0 \end{pmatrix}, \quad (1a)$$

$$m_{11} = \cos\kappa h - \frac{1}{2}\left(\rho - \frac{1}{\rho}\right)\sin\kappa h, \quad (1b)$$

$$m_{12} = -\frac{1}{2}\left(\rho + \frac{1}{\rho}\right)\sin\kappa h = -m_{21}, \quad (1c)$$

$$m_{22} = \cos\kappa h + \frac{1}{2}\left(\rho - \frac{1}{\rho}\right)\sin\kappa h, \quad (1d)$$

$$\rho_{TE} = \frac{\mu_1}{\mu_2}\frac{\kappa h}{\gamma h}, \quad \rho_{TM} = \frac{\varepsilon_1}{\varepsilon_2}\frac{\kappa h}{\gamma h}. \quad (1e)$$

For the SMWG, the modal equation reads $m_{11} = 0$, whereas for the 3MWG, we get $M_{11} = 0$, with:

$$[\mathrm{M}] = [\mathrm{m}]\begin{pmatrix} e^{\gamma d} & 0 \\ 0 & e^{-\gamma d} \end{pmatrix}[\mathrm{m}]\begin{pmatrix} e^{\gamma d} & 0 \\ 0 & e^{-\gamma d} \end{pmatrix}[\mathrm{m}]. \quad (2)$$

By solving the modal equation, as schematically depicted in Fig. 2, we obtain the quantized values of κh; hence the values for the longitudinal wavevector β.

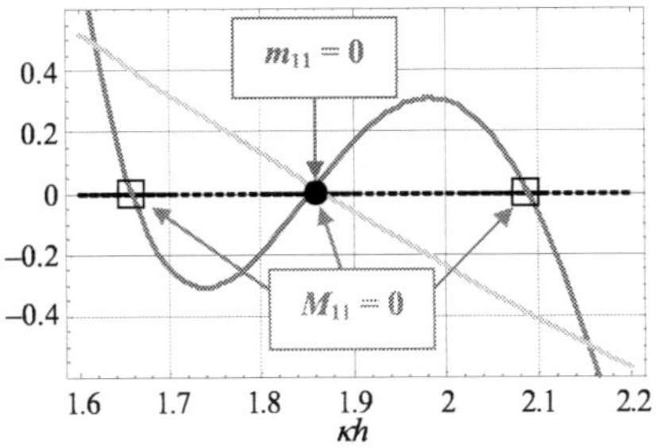

Fig. 2. Graphic depiction of the solution(s) for the TE modal equation, both for the SMWG ($m_{11} = 0$) and the 3MWG ($M_{11} = 0$), in the case where $d = h/2$.

The three super-modes of the 3MWG are a *slow* (fundamental) one, a "*neutral*" one (1st-order) and a *fast* one (2nd-order), of respective wave-vectors (β_S, β_N, β_F).

2.2 Mode profile

For several values of the gap d, we report in Fig. 3 the mode profile $\psi(x) = \psi^+ + \psi^-$, normalized so that:

$$\int |\psi(x)|^2 \, dx = 1. \quad (3)$$

For a vanishing gap ($d = 0$), the super-modes merge continuously into the fundamental, 1st-order and 2nd-order modes of a unique 3MWG of width $3h$.

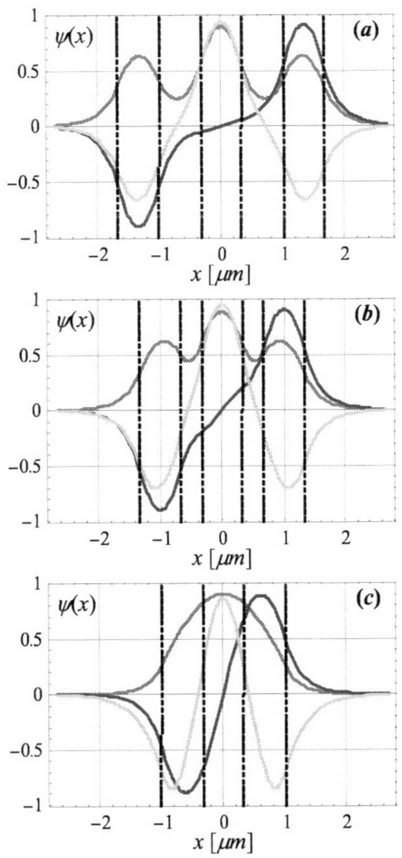

Fig. 3. TE mode profile (in complex amplitude) of the slow, "neutral" and fast super-modes of the ternary slab system: (a) $d = h$; (b) $d = h/2$; (c) $d = 0$.

3 Coupled-Mode Theory

3.1. The simplest model

The simplest model for the symmetric ternary slab system is based on intuitive, realistic assumptions: three identical SMWG, of propagation constant β, are similarly coupled to their nearest neighbour(s) through a real-valued coupling constant χ. Instead of the three fields (F_1, F_2, F_3), let us define the Slowly-Varying Envelope Amplitudes A_n such as $F_n = A_n \, e^{-i\beta z}$; these obey the evolution equation:

$$i\frac{\partial}{\partial z}\begin{pmatrix} A_1 \\ A_2 \\ A_3 \end{pmatrix} = \chi \begin{pmatrix} 0 & 1 & 0 \\ 1 & 0 & 1 \\ 0 & 1 & 0 \end{pmatrix}\begin{pmatrix} A_1 \\ A_2 \\ A_3 \end{pmatrix}. \quad (4)$$

By submitting the reduced operator to a standard diagonalization procedure, we easily derive the eigen-value spectrum to find out that $\beta_F = \beta - \chi\sqrt{2}$, $\beta_N = \beta$, and $\beta_S = \beta + \chi\sqrt{2}$.

Within this simplest model, the wavevectors of the three super-modes appear equidistant in *k*-space, which does *not* coincide with their actual distribution, as determined by an exact matrix analysis: (at least) one more degree of freedom is required.

3.2 Corrective assumptions

The first correction that comes to mind is a possible mismatch affecting the central waveguide, under the influence of its immediate surroundings [Fig. 4].

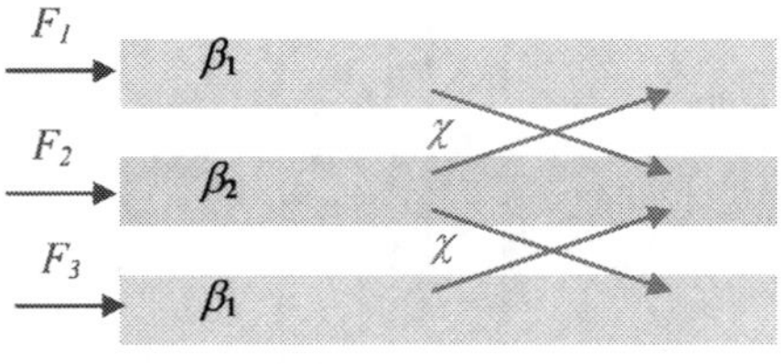

Fig. 4. Ternary slab system with central mismatch.

In this "diagonal perturbation" scheme, we have to consider three slightly different waveguides. The average wavevector β and the mismatch Δ are:

$$\beta = \frac{2\beta_1 + \beta_2}{3}, \quad \Delta = \frac{\beta_2 - \beta_1}{3}. \tag{5}$$

With $\delta = \Delta/\chi$, the new evolution equation is:

$$i\frac{\partial}{\partial z}\begin{pmatrix} A_1 \\ A_2 \\ A_3 \end{pmatrix} = \chi \begin{pmatrix} -\delta & 1 & 0 \\ 1 & 2\delta & 1 \\ 0 & 1 & -\delta \end{pmatrix}\begin{pmatrix} A_1 \\ A_2 \\ A_3 \end{pmatrix}. \tag{6}$$

The calculation of the mismatch-dependent eigenvalues is straightforward: with $R = 8 + 9\,\delta^2$, we get, for any $X \in \{F, N, S\}$, $\beta_X = \beta + \chi\,\lambda_X$, where

$$\lambda_F = \frac{\delta - \sqrt{R}}{2}, \quad \lambda_N = -\delta, \quad \lambda_S = \frac{\delta + \sqrt{R}}{2}. \tag{7}$$

Within this corrective model, the only knowledge of (β_F, β_N, β_S) enables one to proceed to a complete identification of (β, χ, Δ), since $\beta = (\beta_F + \beta_N + \beta_S)/3$ and

$$\chi = \sqrt{\frac{(\beta_S - \beta_N)(\beta_N - \beta_S)}{2}}, \quad \Delta = \beta - \beta_N. \tag{8}$$

Unfortunately, another (fundamentally different) corrective assumption also ensures the missing degree of freedom: that of the "auxiliary coupling" [Fig. 5]:

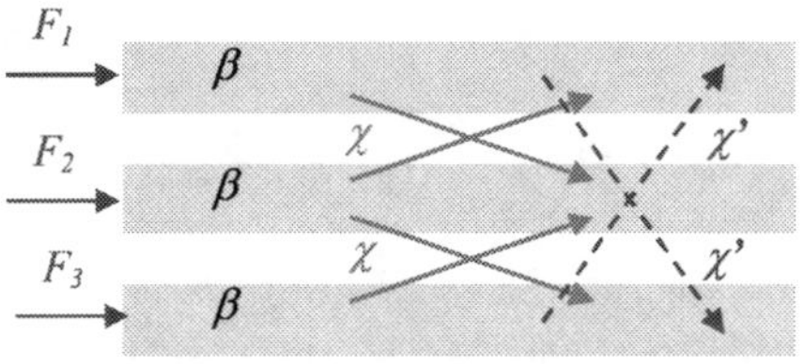

Fig. 5. Ternary phase-matched slab system with auxiliary coupling χ' (with $|\chi'| << \chi$).

In this scheme, the waveguides are again identical, but there is an extra coupling χ' between F_1 and F_3. With $\varepsilon = \chi'/\chi$, the new evolution equation is:

$$i\frac{\partial}{\partial z}\begin{pmatrix} A_1 \\ A_2 \\ A_3 \end{pmatrix} = \chi \begin{pmatrix} 0 & 1 & \varepsilon \\ 1 & 0 & 1 \\ \varepsilon & 1 & 0 \end{pmatrix}\begin{pmatrix} A_1 \\ A_2 \\ A_3 \end{pmatrix}. \tag{9}$$

The calculation of the coupling-dependent eigenvalues is straightforward: with $P = 8 + \varepsilon^2$, we get, for any $X \in \{F, N, S\}$, $\beta_X = \beta + \chi\,\lambda_X$, where

$$\lambda_F = \frac{\delta - \sqrt{P}}{2}, \quad \lambda_N = -\delta, \quad \lambda_S = \frac{\delta + \sqrt{P}}{2}. \tag{10}$$

Within this alternative corrective model, the only knowledge of (β_F, β_N, β_S) enables one – once again – to proceed to a complete identification of (β, χ, Δ), since $\beta = (\beta_F + \beta_N + \beta_S)/3$, $\Delta = \beta - \beta_N$, and

$$\chi = \sqrt{\frac{(2\beta_S - \beta_N - \beta_F)(\beta_S + \beta_N - 2\beta_F)}{18}}. \tag{11}$$

But there is no obvious reason to choose between the two possible corrections: the modal equation doesn't provide enough information to settle the matter.

4 Conclusions

In the frame of Coupled-Mode Theory, we can exploit the properties of the super-modes in order to derive the formal identity between a 3-mode waveguide and a set of three mutually coupled single-mode waveguides. On the "textbook" example of slab systems, we show that the simplest model for a ternary coupler, although based on intuitive realistic assumptions, is incomplete. We explore two different corrective hypotheses that lead to similar results in terms of mode indices. This means that, if a strict term-to-term equivalence is actually legitimate, the solution is not unique. We are currently investigating another criterion: the super-mode profile.

This work is supported by the Labex CominLabs (French National Research Agency program "Investing for the Future" ANR-10-LABX-07-01) within the "3D-Optical-ManyCores" Project (http://www.3d-opt-many-cores.cominlabs.ueb.eu/). The author would also like to thank Prof. Patrice Féron and Dr. Alberto Parini (FOTON Sys.Phot.) for helpful discussions.

References

1. D. Dai, J.E. Bowers, Nanophotonics, **3**, 4-5, 283-311 (2014)
2. Y.G. Boucher, Opt. Eng., **53**,7, 071810 (2014)
3. A. Yariv, P. Yeh, *Optical Waves in Crystals* (Wiley, New York, 1984)
4. T. Tamir (Ed.), *Guided-Wave Optoelectronics* (Springer-Verlag, Berlin, 1990)
5. C.-L. Chen, *Foundations for Guided-Wave Optics* (Wiley, Hoboken, 2007)

Emergence of stable two-colour states in mutually delay-coupled lasers

Masoud Seifikar[1,2,⋆], *Andreas* Amann[1,3,⋆⋆], and *Frank H.* Peters[1,2,⋆⋆⋆]

[1] *Tyndall National Institute, Lee Maltings, Dyke Parade, Cork T12 R5CP, Ireland*
[2] *Department of Physics, University College Cork, Cork, Ireland*
[3] *School of Mathematical Sciences, University College Cork, Cork T12 XF62, Ireland*

Abstract. We theoretically investigate a setup of two mutually delay-coupled semiconductor lasers in a face to face configuration, and study the multi-stabilities and symmetry-broken one-colour and two-colour states for this system, for the development of mutually coupled lasers for integration in a Photonic Integrated Circuit (PIC). We show that the stable two-colour exist for the finite distance between lasers.

1 Introduction

A system of two mutually coupled semiconductor lasers has been widely studied, both experimentally and theoretically [1, 2]. This system is attracting interest for its potential performance in future all-optical signal processing.

In this paper the physics of mutually injection locked lasers within a photonic integrated circuit (PIC) will be investigated. It is of importance in the creation of advanced modulation formats within a PIC and also has additional academic interest in terms of laser dynamics.

The lasers are coherently coupled via their optical fields, where the delay time τ arises from the finite propagation time of the light from one laser to the other. This system, which is illustrated in Fig. 1, is well described by single mode rate equations, which are a system of delay differential equations (DDEs) with one fixed delay. Yanchuk *et al.* [3] predicted one-colour symmetric states for very small delays. Later Erzgräber et al. [4, 5] studied the bifurcations of one-colour states for large delay. Moreover, for zero delay, stable symmetric and symmetry-broken one-colour and two-colour states have been recently predicted by Clerkin et al. [6]. In particular, symmetry-broken two-colour states are highly interesting from an application point of view, for example in the context of all-optical switching [6].

investigate the bifurcation and stability of this system for a finite delay time around $\tau = 1$ (in units of photon lifetime), which means the separation around 1.5 mm for our lasers. Here we show that stable two-colour states continue to exist for finite delays. We perform a detailed study of the relevant bifurcations of the system using the continuation Matlab package DDE-BIFTOOL [7]. As we are interested in the coupling between lasers in a PIC, in this paper we

⋆e-mail: masoud.seifikar@tyndall.ie
⋆⋆e-mail: a.amann@ucc.ie
⋆⋆⋆e-mail: frank.peters@tyndall.ie

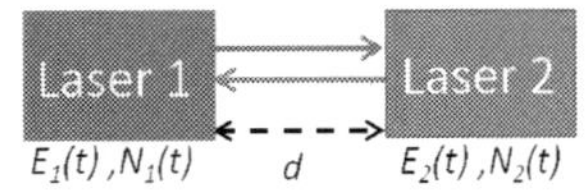

Figure 1. Schematic of two semiconductor lasers, which are mutually coupled by their optical fields.

We present in the next section an overview of the rate equation model that we use to study the physics of the system of mutually coupled lasers. We then solve these DDEs numerically for a given delay, coupling strength and coupling phase in § 3, when we see symmetric and symmetry-broken one-colour and two-colour states for different input parameters. Then in order to understand the origin of these states, we use DDE-BIFTOOL to study the relevant bifurcations of the system with finite delays in § 4, where we first introduce the compound laser modes (CLMs), and explain how CLMs can be continued numerically. We then present the results for symmetric and symmetry-broken states in § 4.1 and § 4.2, respectively. Finally we summarize our conclusion in § 5.

2 Rate equation model

We consider two identical semiconductor lasers in a face to face configuration, as illustrated in Fig. 1. We employ the Lang-Kobayashi-type rate equation to model the two mutually coupled lasers. We write the equation for the normalized complex slowly varying envelope of the optical fields E_1 and E_2, where $E_{1,2} = \Re[E_{1,2}] + \Im[E_{1,2}]$, and the normalized inversions N_1 and N_2, for laser 1 and laser 2, respectively as [5]

$$\frac{dE_1(t)}{dt} = (1+i\alpha)N_1(t)E_1(t) + \kappa e^{-iC_p}E_2(t-\tau) - i\Delta_1(t), \quad (1)$$

$$\frac{dE_2(t)}{dt} = (1+i\alpha)N_2(t)E_2(t) + \kappa e^{-iC_p}E_1(t-\tau) - i\Delta_2(t), \quad (2)$$

$$T\frac{\mathrm{d}N_1(t)}{\mathrm{d}t} = P - N_1(t) - (1 + 2N_1(t))|E_1(t)|^2, \quad (3)$$

$$T\frac{\mathrm{d}N_2(t)}{\mathrm{d}t} = P - N_2(t) - (1 + 2N_2(t))|E_2(t)|^2, \quad (4)$$

Here, time t is measured in the units of photon lifetime, τ_p. T, α and P are the parameters of the Laser. $T = 392$ is the normalized carrier lifetime, calculated by the ratio of the photonic and carrier lifetime. $\alpha = 2.6$ is the linewidth enhancement factor, and $P = 0.23$ is the pumping parameters, which describes the amount of the electrical current used to pump the electron-hole pairs in each lasers. In this paper we concentrate on the case with zero detuning $\Delta = 0$.

Equations 1-4 are given in the reference frame of rotation with the averaged optical angular frequency of the two lasers $\Omega_0 = \Omega_1 + \Omega_2$, where Ω_1 and Ω_2 are the optical angular frequencies of Laser 1 and 2, respectively. Therefore the optical fields of laser 1 and 2 are given by $E_1(t)\exp(i\Omega_0 t)$ and $E_2(t)\exp(i\Omega_0 t)$.

The time delay comes from the distance between two lasers, $\tau = d/c$, where d is the separation of lasers and c denotes the speed of light. As we are interested in the coupling between lasers in a PIC, our main focus is on the small separation around 1 mm. Our main bifurcation parameters are coupling rate κ and coupling phase C_p. Coupling phase can be engineered by changing the distance between lasers as: $C_p = 2 * \pi(d \bmod \lambda_0)$. In the reference frame of rate equations 1-4, C_p is changing with optical frequency as $C_p = \omega_0\tau$. Thus, C_p can be fine-tuned by exploiting the temperature or pump current dependency of ω_0. However, to allow the change interference condition between the optical fields of the lasers, we consider the coupling phase as an independent parameter.

Equations 1-4 are the system of DDEs with a single fix delay, with number of symmetries; S^1-symmetry which is a phase space symmetry [5]. It means that any solution of the rate Eqs. 1-4 is invariant under any phase shift of both electric fields $E_1 \rightarrow E_1 e^{ib}$. Moreover, the system has the π-translational and 2π-translational symmetries. The first one means that π shift in coupling phase C_p is equivalent to a sign change in the optical field of one laser. The second one means that a 2π shift in the phase shift leaves the system unchanged. This means that phase is changing in a circle.

In the rest of this paper we first solve the rate equations 1-4 for given τ, κ and C_p, in § 3. The calculated optical fields frequencies indicates the existence of one-colour and two-colour, symmetric and symmetry-broken states for different values of parameters. In § 4 we then find a CLM of the laser and continue that solution using the continuation software package DDE-BIFTOOL, which helps us to understand the behaviour of the mutually coupled lasers, and predict the stability regions for wide range of given parameters.

3 Solution of the DDEs

In order to solve Eqs. 1-4 we first separate the real and imaginary parts of $E_{1,2}$, which gives a system of 6 DDEs

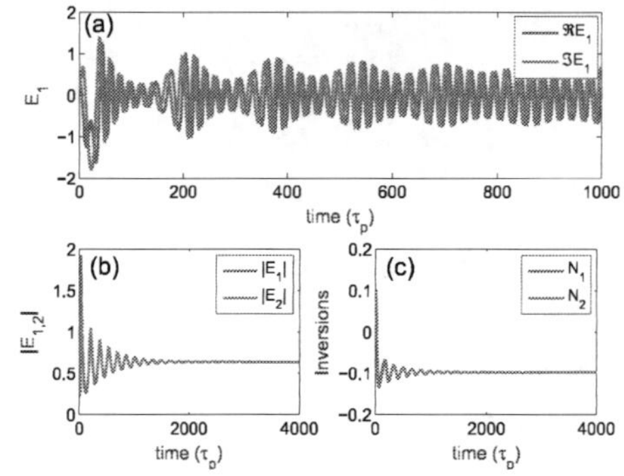

Figure 2. (a) The real and imaginary part of the optical field of the Laser 1. (b) the absolute value of the optical fields and (b) the inversion of Laser 1 and 2, for $\tau = 0.5$, $\kappa = 0.1$ and $C_p = 0.1\pi$.

with a constant delay τ

$$\begin{aligned}
\frac{d\Re(E_1(t))}{dt} &= N_1(t)\Re(E_1(t)) - \alpha N_1(t)\Im(E_1(t)) + \kappa\cos(C_p) \\
&\quad \times\Re(E_2(t-\tau)) + \kappa\sin(C_p)\Im(E_2(t-\tau)), \\
\frac{d\Re(E_2(t))}{dt} &= \alpha N_1(t)\Re(E_1(t)) + N_1(t)\Im(E_1(t)) + \kappa\cos(C_p) \\
&\quad \times\Im(E_2(t-\tau)) - \kappa\sin(C_p)\Re(E_2(t-\tau)), \\
\frac{d\Re(E_1(t))}{dt} &= N_2(t)\Re(E_2(t)) - \alpha N_2(t)\Im(E_2(t)) + \kappa\cos(C_p) \\
&\quad \times\Re(E_1(t-\tau)) + \kappa\sin(C_p)\Im(E_1(t-\tau)), \\
\frac{d\Im(E_2(t))}{dt} &= \alpha N_2(t)\Re(E_2(t)) + N_2(t)\Im(E_2(t)) + \kappa\cos(C_p) \\
&\quad \times\Im(E_1(t-\tau)) - \kappa\sin(C_p)\Re(E_1(t-\tau)), \\
\frac{dN_1(t)}{dt} &= \frac{P}{T} - \frac{N_1(t)}{T} \\
&\quad - \frac{1+2N_1(t)}{T}\left([\Re(E_1(t))]^2 + [\Im(E_1(t))]^2\right), \\
\frac{dN_2(t)}{dt} &= \frac{P}{T} - \frac{N_2(t)}{T} \\
&\quad - \frac{1+2N_2(t)}{T}\left([\Re(E_2(t))]^2 + [\Im(E_2(t))]^2\right),
\end{aligned} \quad (5)$$

Here we have assumed $\Delta = 0$. For given parameters κ and C_p we solve these system of DDEs numerically using Matlab code based on explicit Runge-Kutta method and calculate the optical fields $E_{1,2}$ and inversions $N_{1,2}$. For instance, Fig. 2 displays optical fields and inversions for $\tau = 0.5$, $\kappa = 0.1$ and $C_p = 0.1\pi$.

The time traces of real and imaginary parts of the optical field of Laser 1 is shown Fig. 2.a. Also Figs. 2.b and c indicate that for specific parameters, optical fields and inversions of both lasers converge to constant values after

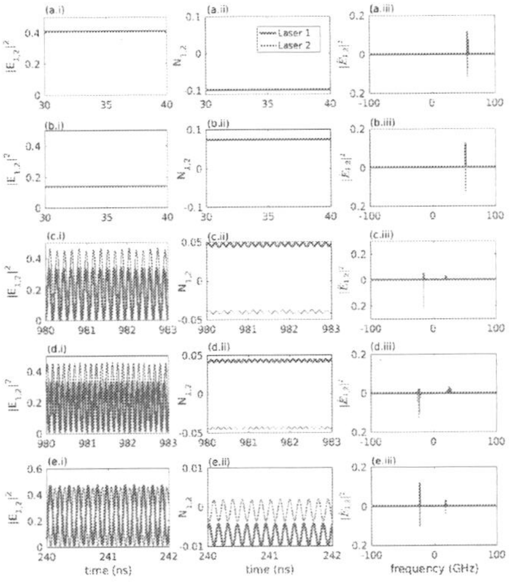

Figure 3. Time trace (left panels), Inversions (middle panels) and frequency spectra (right panels) of Laser 1 and 2, for $\tau = 0.5$, $\kappa = 0.1$ and $C_p = 0.1\pi$ (a panels), $C_p = 0.2\pi$ (b and c panels), $C_p = 0.3\pi$ (d panels) and $C_p = 0.513\pi$ (e panels). Initial conditions for panel a,b, d and e: $\Re(E_1(t=0)) = 0.4$, $\Im(E_1(t=0)) = 0$, $\Re(E_2(t=0)) = -0.4$, $\Im(E_2(t=0)) = 0$, $N_1(t=0) = 0.05$ and $N_2(t=0) = 0.05$, for panel c we only changed $\Re(E_2(t=0)) = -0.39$, and $N_2(t=0) = -0.05$.

about $2000\tau_p$. In these plots time is given in the unit of photon lifetime τ_p, which expresses the rate at which photons are lost from the cavity. For a laser with the cavity length L and mirror reflectivity R_1 and R_2 we can estimate the photon life time by

$$\tau_p = -\frac{2nL}{c}\frac{1}{Ln(R_1R_2)} \tag{6}$$

where n is the refractive index, and c is the light velocity. For instance for the laser cavity length $L = 1.1$ mm, with the low reflectivities $R_1 = R_2 = 0.1$, the photon lifetime is about $\tau_p = 4.9$ ps.

Having the slowly varying optical fields of the lasers we can find the rapidly oscillating physical fields $\tilde{E_1}$ and $\tilde{E_2}$ by taking the Fourier transform of the time trace of the optical fields [6]. Fig. 3 demonstrates the changes in optical field, inversion and frequency of the lasers by changing coupling phase. The time trace of the absolute value of optical fields for $\tau = 0.5$, $\kappa = 0.1$ and several values of C_p are shown in Figs. 3 (a-e).i (the left column panels), and the middle panels (Figs 3 (a-e).ii) display the inversion for given parameters. The frequency spectra of lasers (relative to the optical frequency of the free running lasers) are given in the right hand side plots (Figs 3 (a-e).iii).

In Figs 3.a, we observe a symmetric one-colour state for $\tau = 0.5$, $\kappa = 0.1$ and $C_p = 0.1\pi$, as the lasers have the same optical fields and inversions, and are lasing with a common locked frequency around 56 GHz. By increasing phase to $C_p = 0.2\pi$, we can obtain either symmetric one-colour (Figs 3.b) or symmetry-broken two-colour (Figs 3.c), depending on the initial conditions that is used to solve the DDEs in Eqs. 1-4. However, if we increase phase further to $C_p = 0.3\pi$ we can get only symmetry-broken two-colour stable states, as shown in Figs 3.d. For a very small window of C_p we are able to get symmetric two-colour states as well. For instance, Figs. 3.e display such a states for $C_p = 0.513\pi$.

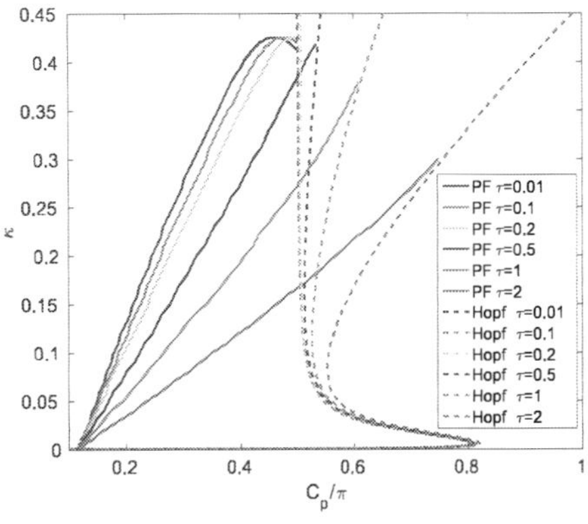

Figure 4. Pitchfork (solid lines) and Hopf (dashed lines) bifurcation diagram of system for $\tau = 0.01 - 2$.

In order to predict the stability of the states for a given parameters, we need to calculate the bifurcation diagram using the DDEs given in Eqs. 1-4. In the next section we use MATLAB package DDE-BIFTOOL [7], which allows us to find the stability information, and bifurcations by following equilibria and periodic orbits.

4 Bifurcation diagrams

Here, we perform a detailed study of the relevant bifurcations of the system using the continuation Matlab package DDE-BIFTOOL. The basic solutions of the system, for a symmetric one-colour state, are given by [5]

$$\begin{aligned} E_1(t) &= R_1^s e^{i\omega_s t}, \\ E_2(t) &= R_1^s e^{i\omega_s t + i\sigma}, \\ N_1(t) &= N_1^s, \\ N_2(t) &= N_2^s \end{aligned} \tag{7}$$

These solutions are called the compound laser modes (CLMs). Here $R_{1,2}^s$, $N_{1,2}^s$, ω_s, and σ are real and time-independent values. We assumed lasers have the same frequency ω_s, which is the deviation between the average solitary laser frequency Ω_0 and the frequency of the coupled laser system. The relationship between the frequency ω_s and the phase difference σ, can be obtained by inserting the ansatz 7 into rate equations 1-4, and eliminating the unknown variables R_1,R_2,N_1,N_2, which gives:

$$\omega_s = \pm\kappa\sqrt{1+\alpha^2}\sin(C_p + \omega_s\tau + arctan(\alpha)) \tag{8}$$

DOI: 10.1051/epjconf/201713900010

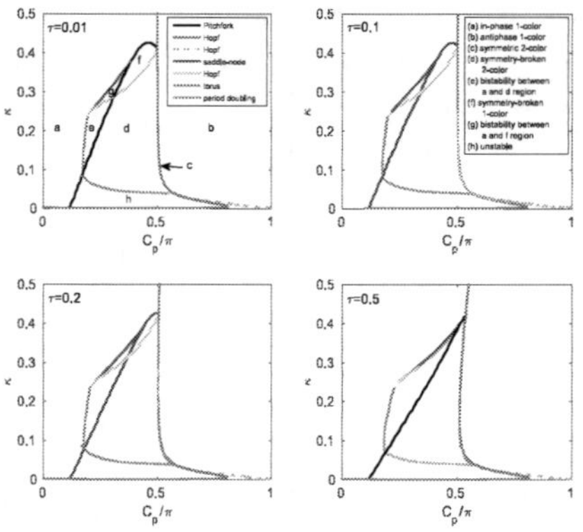

Figure 5. Bifurcations and stability regions in coupling strength κ and coupling phase C_p surface for $\tau = 0.01, 0.1, 0.2$ and 0.5.

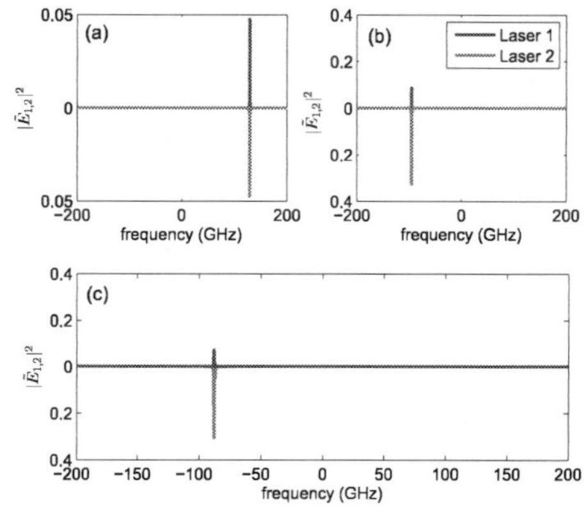

Figure 6. The frequency spectrum relative to the frequency of the free-running lasers, for $\tau = 0.01$, $\kappa = 0.33$ and $C_p = 0.35\pi$ (a and b) and $C_p = 0.35\pi$ (c). Initial conditions for panel a and c: $\Re(E_1(t = 0)) = 0.4$, $\Im(E_1(t = 0)) = 0$, $\Re(E_2(t = 0)) = -0.4$, $\Im(E_2(t = 0)) = 0$, $N_1(t = 0) = 0.05$ and $N_2(t = 0) = 0.05$, for panel b we only changed $\Re(E_2(t = 0)) = -0.39$, and $N_2(t = 0) = -0.05$.

Here + and − are corresponding to the in-phase ($\sigma = 0$) and the anti-phase ($\sigma = \pi$) CLMs, respectively. For a given delay time, we start with the basic solution of the system, given by Eq. 8 and then continue this solution to find bifurcations and stabilities for two bifurcation parameters of coupling phase C_p and coupling strength κ.

4.1 Symmetric solutions

The solution of system can be obtained by the DDE-BIFTOOL, which gives the stability information and bifurcation along each branch. The dashed lines in Fig. 4 show the Hopf lines that are obtained by continuation of the initial Hopf point. The solid lines in Fig. 4 display the pitchfork bifurcations, for delay times $\tau = 0.01 - 2$. The blue curves in this plot show the result for very small delay time ($\tau = 0.01$), which is similar to what Yanchuk [3] has already obtained analytically for $\tau = 0$. The area on the left hand side of the pitchfork line, with the low coupling phase, gives stable in-phase one-colour states, while the region on the right hand side of the Hopf lines, is anti-phase one colour states. At very high coupling strengths, these regions are separated by two supercritical Hopf lines in close proximity. There is no stable symmetric state in the area between Hopf and pitchfork lines, However as we will see in the next section, there are stable symmetry-broken states in this region. Fig. 4 shows that the slope of pitchfork line decreases with increasing delay time τ. The pitchfork-Hopf point is moving to lower κ and higher C_p by increasing the delay time.

4.2 Symmetry-broken solutions

The bifurcation diagram shown in Fig. 4 provides the stability information only for the symmetric states. However using the DDE-BIFTOOL we are able to calculate the symmetry-broken bifurcations as shown in Fig. 5 [4]. Bifurcations and stability regions with the two bifurcation parameters, coupling strength κ and coupling phase C_p are shown in Fig. 5 for $\tau = 0.01, 0.1, 0.2$ and 0.5. The symmetric pitchfork and Hopf bifurcations are shown by black, red and dashed yellow lines. Two Hopf lines are very close for $\kappa > 0.05$, however there is a tiny gap between them which gets wider by increasing delay time, τ. The blue and green curves in Fig. 5 display symmetric-broken saddle-node and Hopf lines, pink line shows the torus bifurcation, and orange line represents the period-doubling bifurcation. The plot for $\tau = 0.01$, shown on the top left panel of this figure is very similar to what Clerkin et al. [6] obtained for zero delay time, using the continuation software AUTO [8].

The bifurcation diagrams in Fig. 5 confirm that we can have stable two-colour states for finite delay time as well. In region d which is confined by pitchfork and Hopf lines at the sides and another Hopf and period-doubling lines at the top and bottom, we observe stable symmetric-broken two-colour states. In region e we can get either symmetric one-colour or symmetry-broken two-colour states as already shown in Figs. 3.b and c. In the tiny region between two Hopf lines (c), we can get symmetric two-colour states as was displayed in Figs. 3.e. Moreover, in region f symmetry-broken one-colour states are stable. In region g we might get symmetric or symmetry-broken one-colour states depending on the initial conditions. This is consistence with Fig. 6, for $\tau = 0.1$ and $\kappa = 0.33$. The corresponding points in Fig. 6.a and b, with $C_p = 0.35\pi$ are located in region g, while Fig. 6.c with $C_p = 0.38\pi$ is located in region f. Fig. 5 indicates that regions f and g are shrinking by increasing the distance between lasers and disappear for τ larger than 0.5.

As shown in Fig. 7, for $\tau = 1$, the torus bifurcation continues along the pitchfork line to reach the Hopf lines.

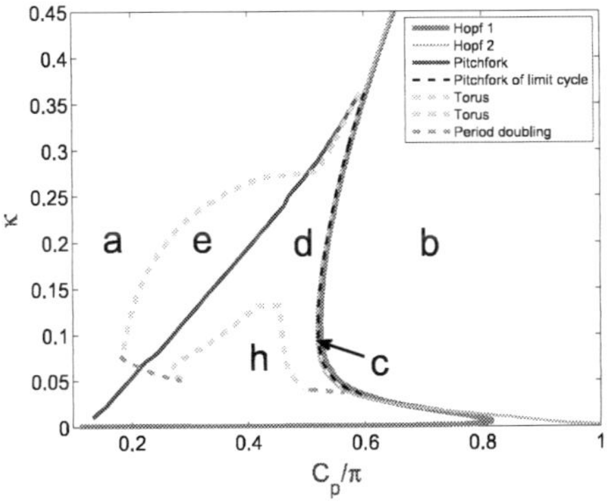

Figure 7. Bifurcations and stability regions in coupling strength κ and coupling phase C_p plane, for delay time $\tau = 1$. Solid lines display the symmetric Hopf and pitchfork bifurcations. The dashed curves show the bifurcations of the limit cycles.Stable symmetric-broken two-coulor states observed for intermediate coupling strength and phase.

Another branch of torus bifurcation emerges at the lower values of coupling which makes the middle part of the period doubling line unstable. Similar to what we have already observed for the lower values of delay time, in regions a and b, in-phase and antiphase one-colour states are stable, respectively. The very small area between two Hopf lines contains region c with stable symmetric two-colour states. More importantly, the model predicts that symmetry-broken two-colour states are stable in the two large regions d and e, which are bounded by torus, period-doubling and pitchfork of limit cycle bifurcations. Region e corresponds to the area of bistability between symmetric one-colour (region a) and symmetry-broken two-colour states (region d).

5 Discussion and conclusions

In summary we theoretically investigated a setup of two mutually delay-coupled semiconductor lasers in a face to face configuration, and studied the multi-stabilities and symmetric and symmetry-broken one-colour and two-colour states for this system. We showed that the stable symmetry-broken two-colour states continue to exist for finite delay time, and therefore stable states exist for almost all values of coupling phase C_p and delay time up to 1. However, increasing delay time makes the area with intermediate C_p and low value of κ unstable. The presented results indicate that while the overall shape of the bifurcation diagram changes rapidly around $\tau = 1$, the stable two-colour states still can be achieved for a wide range of coupling strength κ and coupling phase C_p.

Acknowledgments We thank Bernd Krauskopf and Hartmut Erzgräber for helpful discussions and for help with the DDE-BIFTOOL implementation. This work was supported by the Science Foundation Ireland under grant SFI 13/IA/1960 .

References

[1] H.G. Schuster, K. Lüdge, *Nonlinear laser dynamics: from quantum dots to cryptography* (John Wiley & Sons, 2012)

[2] M.C. Soriano, J. García-Ojalvo, C.R. Mirasso, I. Fischer, Rev. Mod. Phys. **85**, 421 (2013)

[3] S. Yanchuk, L. Recke, Physical Review E **69**, 056221 (2004)

[4] H. Erzgräber, D. Lenstra, B. Krauskopf, Proceedings Symposium IEEE/IEOS Benelux Chapter, 2004, Ghent pp. 147–150 (2004)

[5] H. Erzgräber, B. Krauskopf, D. Lenstra, SIAM Journal on Applied Dynamical Systems **5**, 30 (2006)

[6] E. Clerkin, S. O'Brien, A. Amann, Physical Review E - Statistical, Nonlinear, and Soft Matter Physics (2014)

[7] K. Engelborghs, T. Luzyanina, G. Samaey, TW Report **305** (2000)

[8] E.J. Doedel, T.F. Fairgrieve, B. Sandstede, A.R. Champneys, Y.A. Kuznetsov, X. Wang (2007)

EPJ Web of Conferences **139**, 00011 (2017) DOI: 10.1051/epjconf/201713900011
NANOP 2016

Unified Scattering Parameters formalism in terms of Coupled-Mode Theory for investigating hybrid single-mode/two-mode photonic interconnects

Yann G. Boucher[1-2,*], *Alberto* Parini[1], and *Patrice* Féron[1]

[1]FOTON, Systèmes Photoniques, UMR CNRS 6082, Université de Rennes 1, 22305 Lannion, France
[2]ENIB, Technopôle Brest-Iroise, CS 73862, 29638 Brest, France

Abstract. In terms of Linear Algebra, a directional coupler between a single-mode waveguide and a two-mode waveguide can be thought of as formally equivalent to a set of three mutually coupled single-mode waveguides. Its responses, easily derived in the frame of ternary Coupled-Mode Theory, are used to establish analytically the scattering parameters of a hybrid ring-based modal multiplexer.

1 Introduction

Mode-division multiplexing (MDM), also referred to as Spatial-division multiplexing (SDM), relies on multi-mode waveguides to simultaneously transmit signals allocated on different spatial modes, all sharing the same wavelength [1-3]. In the single-mode domain, the system-level dimensioning of optical multi-port devices is easily carried out by means of the powerful formalism of Scattering Parameters, in which each output wave is simply expressed as a linear combination of all possible input waves. The exchange of energy between modes takes place in directional couplers through an overlap of their evanescent part, a mechanism well described in terms of Coupled-Mode Theory (CMT) [4-5]. There is no fundamental reason why, in the multi-mode domain, this same analytical tool shouldn't appear as relevant.

In what follows, we investigate most specifically the key element for hybrid multimode integrated structures: an asymmetric directional coupler between a single-mode waveguide (SMWG) and a two-mode waveguide (TMWG). Taking advantage of the formal equivalence between a multi-mode waveguide and a set of mutually coupled single-mode ones [6-7], we first derive its responses (transmittance and coupling), before inserting the structure into a more complex system, a ring-based modal multiplexer, for which we establish the scattering parameters under a closed form.

2 Single-mode to multimode coupling

Consider an asymmetric coupler between a SMWG and a TMWG (Fig. 1). The latter is dimensioned in order to support a fundamental guided mode with *even* symmetry and a first-order mode with *odd* symmetry, both characterized by their respective propagation constants β_{even} and β_{odd}. Time dependence is taken as $\exp(+i\,\omega t)$.

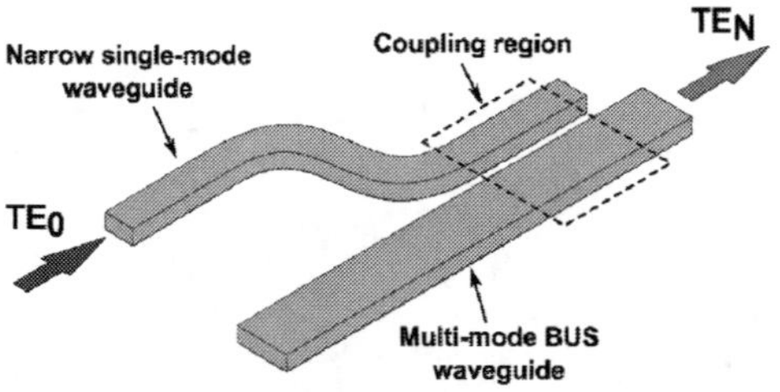

Fig. 1. Typical coupling between a SMWG and a TMWG.

Taking advantage of the formal equivalence between the TMWG and a set of two mutually coupled SMWGs [6], we describe the coupling region as a ternary coupler of length d [7], as schematically depicted in Fig. 2.

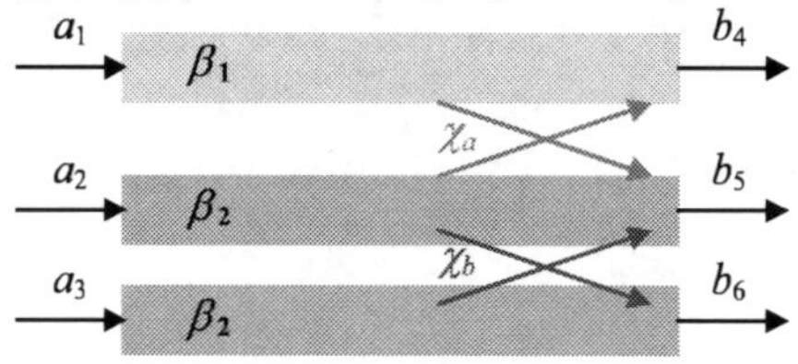

Fig. 2. Modelling of the directional coupler between the upper SMWG and the lower TMWG (itself seen as a set of 2 coupled SMWGs) as a six-port network; $a_i = F_i(0)$, $b_j = F_j(d)$.

In this basis, the evolution operator (along the common z-axis) reads:

$$i\frac{\partial}{\partial z}\begin{pmatrix} F_1 \\ F_2 \\ F_3 \end{pmatrix} = \begin{pmatrix} \beta_1 & \chi_a & 0 \\ \chi_a & \beta_2 & \chi_b \\ 0 & \chi_b & \beta_2 \end{pmatrix}\begin{pmatrix} F_1 \\ F_2 \\ F_3 \end{pmatrix} \qquad (1)$$

where F_n is the complex amplitude of "mode" N° n,

* Corresponding author: yann.boucher@enib.fr

$$\beta_2 = \frac{\beta_{even} + \beta_{odd}}{2}, \quad \chi_b = \frac{\beta_{even} - \beta_{odd}}{2}. \quad (2)$$

The average propagation constant β, the mismatch Δ and the effective coupling constant χ are defined as:

$$\beta = \frac{\beta_1 + 2\beta_2}{3}, \quad \Delta = \frac{\beta_1 - \beta_2}{3}, \quad \chi = \sqrt{\chi_a^2 + \chi_b^2}. \quad (3)$$

Working in terms of Slowly Varying Envelope Amplitudes A_n such as $F_n = A_n\, e^{-i\beta z}$, we get:

$$i\frac{\partial}{\partial z}\begin{pmatrix} A_1 \\ A_2 \\ A_3 \end{pmatrix} = \chi \begin{pmatrix} 2\delta & \sin\alpha & 0 \\ \sin\alpha & -\delta & \cos\alpha \\ 0 & \cos\alpha & -\delta \end{pmatrix}\begin{pmatrix} A_1 \\ A_2 \\ A_3 \end{pmatrix} \quad (4)$$

with $\delta = \Delta/\chi$ the reduced mismatch and $\alpha = \text{atan}(\chi_a/\chi_b)$. Eigenvalues and eigenvectors of the reduced operator are obtained through a standard diagonalization procedure (involving Cardano's method for solving the third-order characteristic equation), leading to a complete determination of the three ternary super-modes (respectively *slow*, "*neutral*" and *fast*). We revert eventually to the even/odd TMWG basis, with:

$$\begin{pmatrix} b_{even} \\ b_{odd} \end{pmatrix} = \frac{1}{\sqrt{2}}\begin{pmatrix} 1 & 1 \\ -1 & 1 \end{pmatrix}\begin{pmatrix} b_5 \\ b_6 \end{pmatrix}, \quad (5a)$$

$$\begin{pmatrix} a_2 \\ a_3 \end{pmatrix} = \frac{1}{\sqrt{2}}\begin{pmatrix} 1 & -1 \\ 1 & 1 \end{pmatrix}\begin{pmatrix} a_{even} \\ a_{odd} \end{pmatrix}, \quad (5b)$$

then we derive analytically the scattering matrix of the coupler, expressed under the following form:

$$\begin{pmatrix} b_4 \\ b_{even} \\ b_{odd} \end{pmatrix} = \begin{pmatrix} t_s & k_{s/e} & k_{s/o} \\ k_{s/e} & t_e & k_{e/o} \\ k_{s/o} & k_{e/o} & t_o \end{pmatrix}\begin{pmatrix} a_1 \\ a_{even} \\ a_{odd} \end{pmatrix}. \quad (6)$$

Because of the symmetries, only six parameters are involved: three "direct transmissions" (t_s, t_e, t_o) and three "cross-couplings" ($k_{s/e}$, $k_{s/o}$, $k_{e/o}$). For the sake of illustration, let us consider the case when perfect phase-matching is achieved between the upper SMWG and the odd mode of the lower TMWG:

$$\beta_1 = \beta_{odd} = \beta_2 - \chi_b \quad \Rightarrow \quad \delta = \frac{-\cos\alpha}{3}. \quad (7)$$

The power exchange between the modes is depicted in Fig. 3 as a function of the dimensionless coupling strength χd, integrated over the interaction length d, when only the upper single-mode (1) is present at the input. Up to 96% of the power can be coupled into the lower odd mode (o). Note that residual coupling into the lower even mode (e) also occurs. For comparison, the dashed line represents the result of a much more simplistic two-mode coupling approach, neglecting the presence of the third (e) mode.

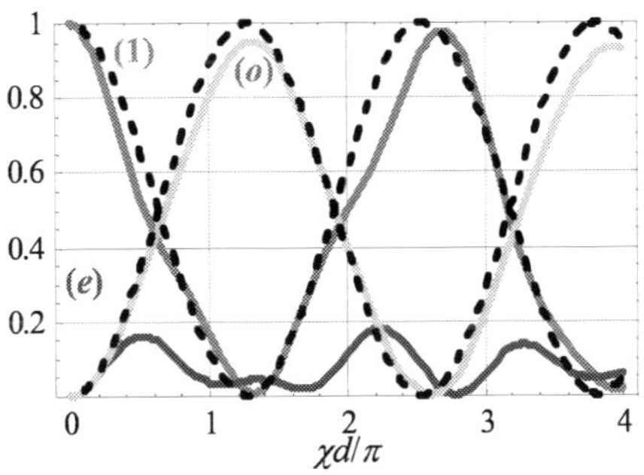

Fig. 3. Power exchange as a function of the coupling from input a_1 to outputs b_1, b_{even}, b_{odd}. In dashed line, the result of a much more simplistic approach with only two-mode coupling, neglecting the residual crosstalk into the mismatched (e) mode.

We would like to emphasize that the calculations are carried out under a reduced form, the only assumption being that of eqn.(7); this means that the results carry a universal significance for any "*single to odd*" phase-matched directional coupler, whatever the precise opto-geometric parameters of the actual device.

Let us now investigate a more elaborate device: a ring-based modal multiplexer.

3 Ring-based modal multiplexer

The system under consideration [Fig. 4] is made of an input SMWG connected to an output TMWG through a ring resonator (RR) in the "racetrack" configuration. For the sake of clarity, let us assume that the SMWG and the RR share the same propagation constant β_1. The upper binary coupler, of length d, is completely defined by its own scattering parameters: direct and cross-transmission t_1 and k_1, respectively. The lower ternary coupler, also of length d, is described as seen above. The whole device works in a way similar to a single-mode add-drop filter, except for the additional feature of mode conversion.

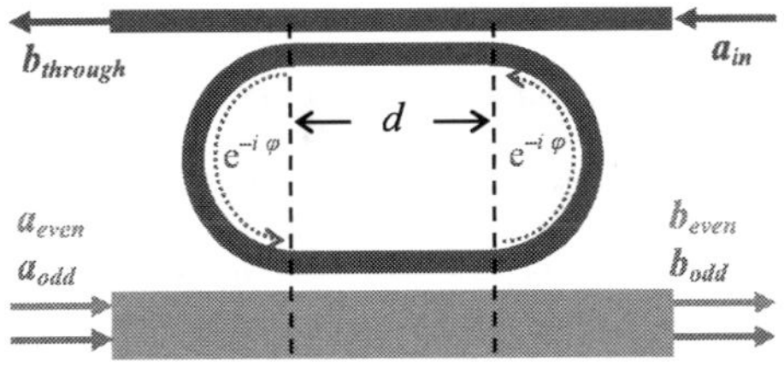

Fig. 4. Ring-based modal multiplexer seen as a multi-port network. All ports (with their phase reference) are actually located at each end of the couplers.

Once again, we take advantage of Linear Algebra in order to express each relevant output field ($b_{through}$, b_{even}, b_{odd}) as a linear combination of all input fields (a_{in}, a_{even}, a_{odd}). Once closed the feedback loop provided by the RR, a straightforward self-consistent calculation leads eventually to the following analytical expressions:

$$b_{even} = a_{in} \frac{k_1 k_{s/e} \, e^{-i\varphi}}{D} + a_{even} \left(t_e + \frac{t_1 k_{s/e}^2 \, e^{-2i\varphi}}{D} \right) + a_{odd} \left(t_o + \frac{t_1 k_{s/o}^2 \, e^{-2i\varphi}}{D} \right), \tag{8a}$$

$$b_{odd} = a_{in} \frac{k_1 k_{s/o} \, e^{-i\varphi}}{D} + a_{even} \left(k_{e/o} + \frac{t_1 \, k_{s/e} \, k_{s/o} \, e^{-2i\varphi}}{D} \right) + a_{odd} \left(t_o + \frac{t_1 \, k_{s/o}^2 \, e^{-2i\varphi}}{D} \right), \tag{8b}$$

$$b_{through} = a_{in} \left(t_1 + \frac{k_1^2 \, t_s \, e^{-2i\varphi}}{D} \right) + a_{even} \left(\frac{k_1 \, k_{s/e} \, e^{-i\varphi}}{D} \right) + a_{odd} \left(\frac{k_1 \, k_{s/o} \, e^{-i\varphi}}{D} \right), \tag{8c}$$

with D the unmistakable signature of a feedback loop:

$$D = 1 - t_1 \, t_s \, e^{-2i\varphi}, \tag{9}$$

where φ denotes the phase change through a half-circle inside the RR (from the end of the upper coupler to the entrance of the lower one, or *vice versa*). The complex quantity $t_1 \, t_s \, e^{-2i\varphi}$ represents the amplitude and phase change over a round-trip, experienced by a wave circulating in the RR.

The overall form taken by scattering parameters is worth some comments. From eqn.(8), one can see that they fall into one of two categories. The simplest ones (from a_{in} to b_{even} or to b_{odd}, from a_{even} or a_{odd} to $b_{through}$) take the form of a single-pass numerator over the common denominator D. The other ones come as the sum of two terms, expressing the interference between two possible paths: for instance, the transmittance from a_{in} to $b_{through}$ results from the competition between the direct path through the upper coupler (t_1), and the path through the RR ($k_1^2 \, t_s \, e^{-2i\varphi}$), itself amplified by the feedback loop (D).

It is therefore possible to exploit these interference effects in order to fine-tune the parameters of the whole system. For instance, by equating the internal transmission coefficients t_1 and t_s on both couplers seen from inside the RR, it becomes possible (by destructive interference) to exactly cancel the transmittance from a_{in} to $b_{through}$, thus converting all the power of the upper single-mode input into the output modes of the lower TMWG. Note, however, that such a mechanism relies on interference: in practice, we have to make sure that the working wavelength is actually attuned to a resonance. Tolerance and bandwidth issues have to be kept in mind, but will not be considered further in the present work.

Besides, it can be interesting to note that, once established the scattering parameters of both couplers, expressions (8), themselves closely related to the topology of the network, could also be recovered from graph theory considerations [8].

4 Conclusions

We have presented a modelling technique based on scattering matrix formalism adapted to multi-mode devices. By exploiting the formal identity between a two-mode waveguide and a set of two mutually coupled single-mode waveguides, the multi-mode interaction is treated within the usual framework of Coupled Mode Theory (CMT). This approach has been tested on the typical example of a ring-based modal multiplexer, opening the way to a systematic exploration of the parameter space for the purpose of design. Taking into account possible losses in the waveguides would be straightforward.

This work is supported by the Labex CominLabs (French National Research Agency program "Investing for the Future" ANR-10-LABX-07-01) within the "3D-Optical-ManyCores" Project (http://www.3d-opt-many-cores.cominlabs.ueb.eu/). The authors would also like to thank Prof. Christophe Peucheret (FOTON Sys.Phot.) for helpful and stimulating discussions.

References

1. Y. Ding, J. Xu, F. Da Ros, B. Huang, H. Ou and C. Peucheret, Opt. Express, **21**, 8, 10376-82 (2013)
2. D. Dai, J.E. Bowers, Nanophotonics, **3**, 4-5, 283-311 (2014)
3. C. Peucheret, Y. Ding, J. Xu, F. Da Ros, A. Parini, H. Ou, Advanced Photonics 2015 OSA Technical Digest (online), paper SpT2E.3 (2015)
4. A. Yariv & P. Yeh, *Optical Waves in Crystals* (Wiley, New York, 1984)
5. T. Tamir (Ed.), *Guided-Wave Optoelectronics* (Springer-Verlag, Berlin, 1990)
6. Y.G. Boucher, Opt. Eng., **53**, 7, 071810 (2014)
7. Y.G. Boucher, A. Parini, P. Féron, Optical Quantum Electron. (to be published)
8. E.J. Henley & R.A. Williams, *Graph Theory in Modern Engineering* (Academic Press, New York and London, 1973)

Linearly and circularly polarized laser photoinduced molecular order in azo dye doped polymer films

Bendaoud Saad*

National School of Applied Sciences of Safi, Cadi Ayyad University, Sidi Bouzid's Road, B P 63, Safi, 46000, Morocco

Abstract. Photo-induced behavior of Azo Disperse one (AZD1) doped Poly(Methyl MethAcrylate) (PMMA) using both linear and circular polarized light is studied. The anisotropy is not erased by the circular polarization light. The circular polarization light combined with relatively long lifetime of the cis state in azo dye doped polymers activate all transverse directions of the angular hole burning through the spot in the film inducing anisotropy. Under circular polarized light, there is no orientation perpendicularly to the helex described by the rotating electric field vector, trans molecules reorients in the propagation direction of the pump beam. The polarization state of the probe beam after propagation through the pumped spot depends strongly on the angle of incidence of both pump and probe beams on the input face. In the case where circular polarized pump and probe beams are under the same angle of incidence, the probe beam "sees" anisotropic film as if it is isotropic. Results of this work shows the possibility to reorient azobenzene-type molecules in two orthogonal directions using alternately linearly and circularly polarized beams.

1 Introduction

Amorphous azo dye doped polymers are very interesting for their potential applications in optics and photonics [1]. These potentialities come from the photo-isomerization that modifies the molecular dielectric polarization and thus the refractive index as well as the absorption of the material. By using polarized light, one can induce collective molecular reorientation in a favored direction that leads to the anisotropy i.e., dichroism and birefringence. The optimization of these properties is important for example for All-Optical Switching [2] and Optical Data Storage [3], since modulations achieved are very high. If azobenzene molecules are excited using a linear polarized light, the angular hole burning occurs for the azo dye's transition dipole moment axis has a component parallel to the electric field (E-field) of the light [4]. An angular redistribution of trans molecules occurs when they relax from cis state to trans state. The molecules that are perpendicular to the E-field of the light are no longer pumped; the photo-isomerization process will tend to accumulate trans isomers in that direction that is the direction perpendicular to the oscillating E-field of the incoming light [5]. Circularly polarized light was used to erase the anisotropy in polymers containing azobenzene groups [6]. This technique of erase was reported for the first time by Rochon et al. [7]. Since then, while the theoretically use of this idea continue [8-14], no significant studies have been published for further verification.

In our previous study [15], a transient out of plane rotation of the director in azo dye doped nematic liquid crystals using linearly polarized light was observed. Although, the molecular reorientation in viscous substances such liquid crystals is completely different from that in solid matrices such PMMA; PS and PVA polymers. In viscous matrices, the reorientation can involve a collective motion of the polymer molecules hosts themselves whereas in solid matrices, the reorientation can involve only the azobenzene-type molecules while long polymeric chains not move. In azo dye doped solid polymer however, the photo-isomerization of azobenzene-type molecules cause significant photo-mechanical deformations in the polymer matrix even at microscopic scale [16].Also, there is a big difference between azobenzene molecules *bonded* to (or in) the polymer chain and the *unbonded* one, hence the classification of these materials into two categories arises: azobenzene groups containing polymers (the copolymers) and azo dye doped polymers, and this is another subject that is outside our current talk.

Unpolarized light can push the azo molecules in viscous polymer containing azos [17] to orient in a direction which will be along the propagation direction of the excitation light. Dumont et al. are reported that using circular polarized light to pump Disperse Red one (DR1) doped in PMMA induces the angular redistribution and produces photo-assisted poling of the azobenzene molecules [18,19]. Barrett et al. are demonstrated that circular polarized light induces a reversible anisotropic dilatation due to the photo-mechanical effect which is due in turn to the photo-isomerization in thin films of azo polymers [20]. Sajti et al. [21] are reported that upon illumination with circularly polarized light, the chromospheres of E1aP12 polymer tend to orient themselves towards the wave vector of the excitation light. Ramanujam et al. [22] are

* Corresponding author: b.saad@uca.ma

published a series of color pictures taken by a polarization microscope revealing that both linear and circular polarized light creates anisotropy in isotropic azobenzene polymers. Although, inducing the anisotropy using a linear polarized light is easy to understand owing to the reorientation of the azos molecules in directions perpendicular to the E-field direction of the excitation light. However, this fails to explain why circularly polarized light induces the anisotropy.

Nevertheless, if the light is circularly polarized, in principle, it's obvious that the E-field plane-of-vibration which rotates at the frequency of the light ($\nu \sim 5\times10^{14}$ Hz) will be prohibited from being occupied by the trans molecules [23]. Hence, it must be emphasized, however, that there is another favored direction during photo-isomerization which will be along the propagation axis of a circular polarization of the incident excitation light. It is unavoidable that azobenzene-type molecules will efficiently buildup aligned along the propagation axis of the circular polarized light but this is often ignored in the literature, or interpreted as a photo-bleaching of the sample. Nunzi et al. are reported that circular polarized light not erase birefringence in cyano azobenzene polymer films, but induces the birefringence [24]; the authors have considered this as due to a phase transition in the studied liquid crystal polymer enhancing order. Chuang et al. [25] are published that the azobenzene molecules can be further aligned by recovering in films two incoherent pump beams of the light with crossed linear polarization states; this enhances the birefringence. In order to generalize the selective angular redistribution to the whole of a given sample, one possible way is trying to irradiate simultaneously several rectilinearly polarization directions. Since E-fields of nonpolarized light exists at each instant in all transverse directions, unpolarized light such power white light sources or unpolarized lasers can push the azo dyes to orient along the axis of the incoming light [26,27], even sunshine is suitable [28,29].

In our best acknowledgement, no work has been published on azo dye doped polymer films using circularly polarized pump and probe light irradiances. We report here the results of the photoinduced behavior of an azo dye doped polymer by using both linearly and circularly polarized light beams. Not only a circular polarized light not erase the photo-induced anisotropy but induces it in this azo dye doped polymer. Unambiguous results of the reorientation of azobenzene molecules towards the propagation direction of the light have been performed.

2 Experimental descriptions

2.1 Fabrication of the films

The azobenzene molecule used in this work is the azo dye called AZD1 [30-33]. The polymer is the poly(methyl methacrylate) abridged as PMMA (average Mw : ~120,000 by GPC powder, Tg=105°C). Products are from Sigma-Aldrich. AZD1 doped in PMMA samples are obtained from the mixture of two separately prepared solutions. The first is the solution of the AZD1 in tetrahydrofuran (THF) solvent. The second one is the PMMA powder dissolved in the same solvent on a magnetic stirrer for 48 hours which was used to ensure the homogeneity of solutions and prevent the formation of eventual aggregates. The gravity deposition technique was used for deposing the solution on the microscope slides (Fisher Scientific Co.) previously cleaned and dried. Next, samples was left for 24 h in the closed, small Plexiglas box to get a regulated solvent evaporation rate allowing the mixture of azo dye and polymer solution to dry slowly on a leveled glass substrates under normal laboratory conditions (20°C, RH=35-45%) during one week in darkness. Films were then heated for 1 h at 100 °C in an oven to remove any remaining solvent [4]. The average thickness measured with a Sloan's Dektak IIA profilometer is 40 μm. Weakly doped films are analyzed with a Varian Cary 100 UV-Vis spectrophotometer, a typical spectrum is depicted in Fig. 1.

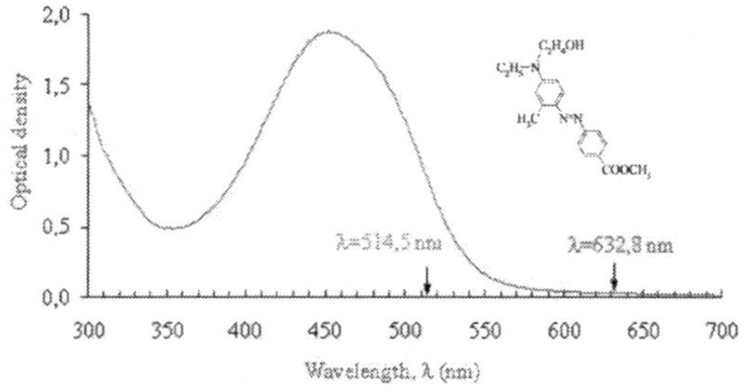

Fig. 1. UV-Vis absorption spectra of AZD1 doped PMMA film (0,5wt%). The inset shows the trans form of AZD1 molecular structure.

2.2 Polarimetric experimental set-up

In the experimental setup cited in many references, some examples are therein [8, 12, 24, 34-37], pump and probe beams are usually linearly polarized and the polarization direction of the probe is oriented at an angle of 45° with respect with that of the pump. Also, the film was set between a pair of crossed polarizer analyzer in the optical path of the probe beam. In general, the incidence angles of pump and probe beam are 0° and 45° respectively (or vice versa). However, this experimental setup is limited at least for three reasons.

First, the use of a rectilinear polarization beam restricts the angular hole burning to one direction; and only a certain percentage of azo molecules are activated. Second, only one transmittance component of the linearly polarized probe beam is detected after the analyzer. This leads to lose so much of data on what is happening in the sample. So change in the anisotropy was often interpreted as "randomization" and "erasure" [2, 35] if the transmittance of the probe beam decreases and as "photo-bleaching" [8, 38] if it increases whereas in reality, as our results will show, there is neither erasure nor photo-bleaching of the sample [21], photo-induced behaviors of polymers containing azos are simply a manifestation of reversible anisotropy changes in various forms introducing phase shift between vertical and horizontal components of the E-field of the probe beam leading to the change in its polarization state. The third reason is it does not allow us to probe the created anisotropy at different angles.

Our experimental set-up is shown in Fig. 2. The surface of the film is in Oxy plane. A continuous wave argon ion laser (λ=514,5 nm, ϕ_e~100 mW) was used as excitation source, and low power He-Ne laser (λ=632,8 nm, ϕ_e~1 mW) being a probe. The two wavelengths are placed on absorption spectrum (Fig. 1). The pump beam was fixed permanently at normal incidence on the input surface of photosensitive thin layer, i.e. parallel to Oz axis as shown in Fig. 2.

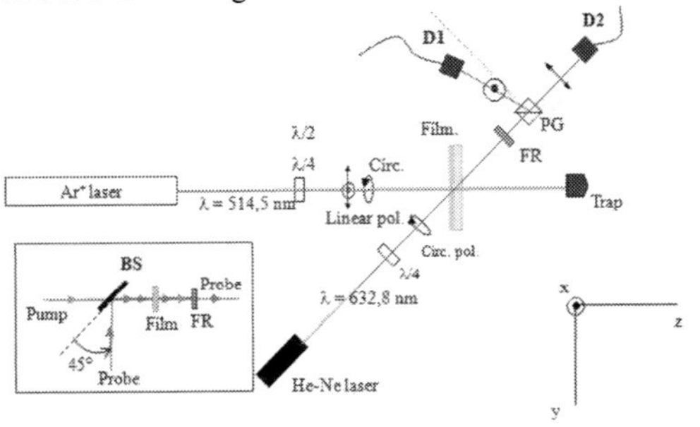

Fig. 2. Polarimetric Experimental setup.
The pump is a linearly polarized continuous wave argon ion laser beam emitting at 514,5 nm. The probe beam is a linearly He–Ne laser at 632,8 nm. Optical components: $\lambda/2$ and $\lambda/4$ are half-wave and quarter-wave plates respectively. The PG component is the Glan-Taylor prism. Optoelectronic devices **D1** and **D2** are detectors of the light. The optical component FR is an interferential filter at 633 nm. Inset the coaxially pomp and probe beam at incidence normal on the film.

The probe beam wave vector located in the Oyz incidence plane is for the moment under an angle of incidence of 45° relative to normal incidence on the film. A linear polarized probe beam is circularly polarized by using a $\lambda/4$-Plate at λ~632,8 nm. The E-field of circular polarization state rotates at the frequency of the light (~$4{,}74\times10^{14}$ Hz) scanning helically all-orthogonal directions to its wave vector especially in the film. A Glan–Taylor prism (Thorlabs Inc., Coating: 350 - 700 nm, extinction ratio: 100,000:1) was used to splits the circularly polarized probe beam into two equal rectilinear polarized components: the S-polarization component oscillates in the direction of the Ox axis and detected by the detector D1 and the P-polarization component in the incidence plane Oyz and detected by the detector D2. The P-component make an azimuth angle φ with the Oy axis of 45°. The waists of the pump and probe beam at the plane of thin layer are approximately $2{,}94\times10^{-3}$ m and $0{,}52\times10^{-3}$ m respectively. Since the probe beam was centered in the illumined area of the film.

3 Results and discussion

3.1 Induced anisotropy using P-polarization

A fresh isotropic film (concentration~2,8wt%, thicknesses ~40 μm) were pumped using P-polarized beam for enough time to reach saturation then switched off. The irradiance is 3,61 kW/m^2 for the pump and ~$1{,}18\times10^{-3}$ kW/m^2 for the probe. To compare, the sunlight irradiance external to the Earth's atmosphere is 1,36 kW/m2. Temporal behaviors of S- and P-components of the probe beam are detected simultaneously. Figure 3 shows a complete cycle of photo-inducing anisotropy and its self-relaxation.

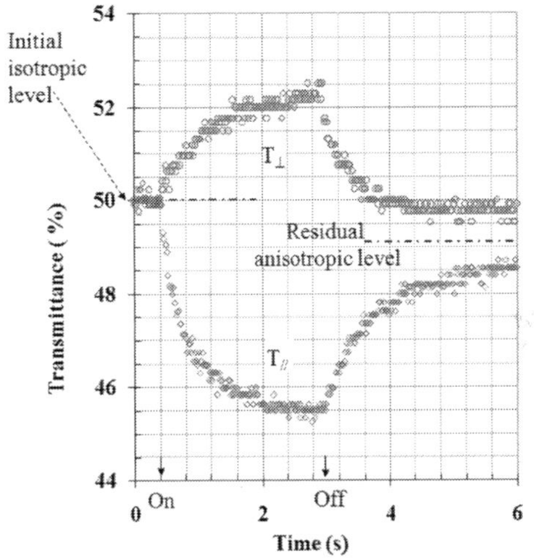

Fig. 3. One cycle of photo-buildup/self-randomization anisotropy in AZD1 doped in PMMA polymer film pumped by using CW Argon ion laser (λ=514,5 nm).

Between the instants t=0 and t=0,4 s, fresh sample is not yet pumped. The $T_\perp$ and $T_{//}$ transmittances corresponding respectively to S- and P-polarization components of the circular polarized probe beam are equal ($T_\perp$=$T_{//}$=50%), indicating the random orientation of the azobenzene molecules [9]. In this case, the refractive indexes n_x, n_y and n_z corresponding to the Ox, Oy, Oz directions (see Fig. 2) are equal i.e., the effective refractive index ellipsoid is spherical. The S- and P-components of the probe beam have in fact "seen" the same refractive index in propagating in their optical paths so the intensities of the S- and P-components are equal at each instant in this time interval.

Between the instants t=0,4 s and t=3 s, the pump light is turned on, the $T_\perp$ component of the probe beam increases while the $T_{//}$ decrease, and the resulting anisotropy increases. The photo-isomerization continue until all the azo molecules which can participate in this mechanism lie in the perpendicular direction, then the anisotropy process is saturated [5]. In this time interval, a negative uniaxial medium was been photo-induced [25]. The optical axis of the medium is towards the Oy axis which is in the same direction of rectilinear P-polarization of the pump beam, and its extraordinary and ordinary refractive indices are n_e=n_y and n_o=n_x≈n_z

respectively. Under pumping, S-polarization component of the circular polarized probe beam "see" the ordinary refractive index changes so the corresponding $T_{\perp}$ transmittance increases whereas the $T_{//}$ transmittance component corresponding to the P-polarization component decreases because it "see" a refractive index $n(\varphi)$ changes given by the instantaneous equation of the index ellipsoid whereabouts $n_e(0)<n(\varphi)<n_o$. Indeed, an oblate ellipsoid of refractive indices whose extraordinary axis is along the Oy axis and the ordinary axis is along the Ox axis or Oz axis or any other axis in the Oxz plane passing through the origin (0,0,0). The evaluated birefringence by using $I=I_0\sin^2(\Delta\varphi/2)$ where $\Delta\varphi=2\pi\Delta nd/\lambda$ [5,39] is $\Delta n=(n_o-n_e)\sim1,21\times10^{-3}$. The time required to achieve saturation takes only few seconds. Note that the transmittance (in absolute value) is higher for P-polarization component of the probe than for its S-polarization one.

At instant t~3 s, pump light is turned off, the spontaneous relaxation starts [18,19]. There is no need to change the polarization state of the pump to randomize the oriented molecules, simply cut-off the light and the randomization starts itself i.e., auto-relaxation [30,31]. The whole inducing and self-randomization cycle is very fast (3-6 s). As one can see, the $T_{\perp}$ component of the probe beam decreases while the $T_{//}$ component increases "exponentially" and the resulting anisotropy decreases. This is due to the thermal reorientation of azobenzene molecules which tends to restore the thermodynamically randomized state [7,40]. Note that the final levels ($T_{\perp}$ and $T_{//}$) of both S- and P-components after switching off the pump tend towards another steady state level that is quite different from the initial level before the first pumping.

Thereafter, the film was cyclically pumped with the same P-linearly polarized beam. Figure 4 shows four inducing/relaxation cycles.

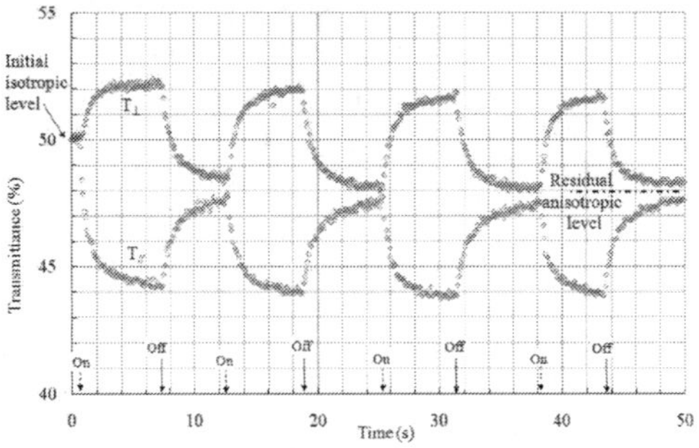

Fig. 4. Four photo-induced/relaxation cycles in AZD1 doped PMMA film cyclically pumped with On/Off P-polarized beam.

The first cycle in Fig. 4 is identical to that shown in Fig. 3. At instant t=13 s in Fig. 4, the P-polarized pump is turned on again on the same region (not in its amorphous state). As one can see, its effect is to produce the same amount of anisotropy in the same time interval like for the first cycle. During the second cycle and so on, the maximum birefringence reached at the saturation is $\Delta n\sim2,01\times10^{-3}$. A difference compared to the first cycle is that the $T_{\perp}$ and $T_{//}$ transmittances are now symmetrical with respect to the horizontal line that indicates "persistent partial anisotropic level". This is because the pumped region is already partially ordered after the first irradiation and is not yet randomized completely. Indeed, at microscopic scale [38], build-up anisotropy again on the same irradiated region involves moving the azo dyes from a thermodynamically stable state with partially anisotropy to another oriented state identical to the previous one [40]. Even the initial isotropic level is never restored by pumping with the same beam whatever the direction of its linear polarization state, the reversible anisotropy, which was attributed to the molecular reorientation induced by the photo-isomerization of the azobenzene molecules, once no irreversible processes [8] that could indicate sample's photo-bleaching were observed.

3.2 Alternately S- and P-polarizations

A fresh film is pumped by using the same alternately S- and P-polarized beam. The results are shown in Fig. 5.

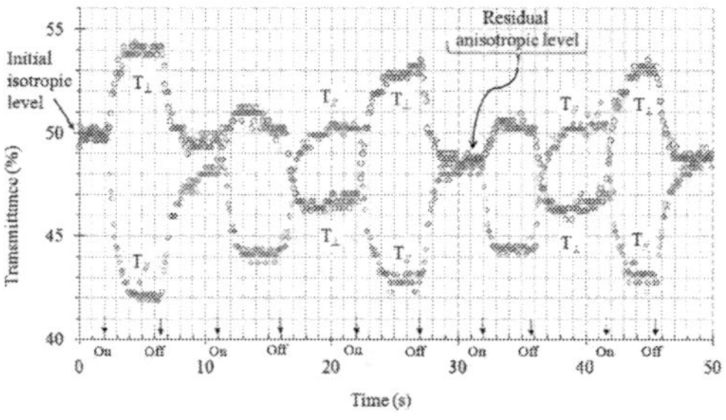

Fig. 5. Four photo-induced/relaxation cycles in a AZD1 doped PMMA film cyclically pumped with successively On/Off S- and P-polarized beam.

Consider for the moment the first cycle. Between instants t=0 and t=2 s, the sample is not yet pumped. At t=2 s, the pump beam is switched on, the angular redistribution occurs subsequent to the photo-isomerization process. Long axis of trans molecules will be gradually oriented perpendicularly to the E-Field of the pump beam i.e., in the Oy direction. The ordinary refractive index $n_o=n_x\approx n_z$ increases to a higher value, while the extraordinary refractive index $n_e=n_y$ decreases to the lower value i.e., $n_e=n_y<n_o=n_x\approx n_z$. Hence, the film does transform from an isotropic to a negative uniaxial medium [25]. One obtains an oblate ellipsoid of refractive indices with its extraordinary axis along the Ox axis and the ordinary axis is along Oy axis. At the saturation, the three principal refractive indices reach their dynamic equilibrium values, thus n_x and n_z are saturated to a larger value while n_y to a smaller value in comparison with those of the isotropic film.

The second cycle in Fig. 5 show that exciting the film by using P-polarized beam immediately after the first irradiation with S-polarization state affect terribly the amplitude of $T_{\perp}$ and $T_{//}$ transmittances. At t~15,5 s, the pump is switched off, $T_{\perp}$ decreases and $T_{//}$ increases.

Surprisingly they cross at t=17 s and $T_\perp$ continue to decrease whereas $T_{//}$ continue to increase towards completely different stable levels that are symmetrical with respect to final stable level reached at instant t=29 to 30 s or just before t=50 s (see Fig 5). After switched off the P-polarized beam, the amount of trans molecules that align in Ox direction continue to increase, and that's why $T_\perp$ continue to decrease and $T_{//}$ to increase. In our best knowledge, the reorientation of the molecule in a proffered direction without light is new phenomenon has never been reported in the literature. This new behavior is probably due to a complex interactions between the azobenzene molecules and the matrix host subsequent to the photomechanical effect [16,40,41].

The third cycle in Fig. 5 is obtained by pumping again using S-polarized. As one can see, the changes in $T_\perp$ and $T_{//}$ probe components are reversible. The cycle starts at instant t~ 22 s. As the pump beam is switched on, azoic molecules reorient perpendicularly to the S-polarization plane. Therefore $T_\perp$ component increases while $T_{//}$ decreases simultaneously, then they cross each other at point t~22,5 s exactly at the residual stable partial anisotropy level. For the excited S-polarization state, the fifth cycle is identical to the third cycle and this latter is not identical to the first one, this because the start point is not the same. For the excited P-polarization, the fourth cycle is identical to the second cycle; and so on. More than hundred identical S- and P-polarized pump cycles have been achieved without the appearance of any signs of fatigue in the sample. It should be noted that the residual level at t=50 s (Fig. 4 and Fig. 5) is displaced downwards compared with the initial level of isotropy. At long-term; $T_\perp$ and $T_{//}$ components tends towards another enduring steady state level which is quite different from the initial isotropic one; we have called it "partial residual anisotropic level" (Fig. 3). This behavior is a common characteristic of azo dye containing polymers films. Nevertheless, the thermal relaxation of molecules back to randomize trans state is incomplete: the anisotropy does not recover its initial isotropic value, either because of an incomplete cis–trans relaxation or because of some deformations of the doped polymer [42]. When the laser is turned off, a substantial number of cis azobenzene isomers remain "trapped" in their new orientation [7,40] and long-term macroscopic anisotropy has been optically.

3.3 Linearly versus circularly polarizations

Three principal polarimetric experimental configurations of interaction are studied. Fresh doped films has been used. Results are presented in Fig 6.

First, the pump beam is P-polarized. Curves labeled 1 correspond to this mode of excitation. Second, the pump is S-polarized. Curves labeled 2 correspond to this mode of excitation. Third, the pump beam is circularly polarized i.e., its E-field rotates helically around the Oz axis at the angular frequency of the light (~5×10^{14} Hz). Curves labeled 3 correspond to this mode. Curves labeled 1(sp) and 1(pp) of the first cycle in Fig. 6 are the same as those shown in Fig. 3. For the instance, a curve 2(ps) corresponds to the P-polarization of the probe and S-polarization of the pump beams, and so on. The probing time interval for the first and second cycles are Δt=2,6 s and Δt=2 s respectively. Curves 2 and 3 consist each of an only one photo-induced/relaxation cycle. For these last, the pumping time interval is longer than this of curves 1. Saturation levels are almost accomplished. The most important finding is that the greatest anisotropy was induced when doped films were pumped by circularly polarized pump beam i.e., the circularly polarized light not erase anisotropy, but favors a certain order of the azoic molecules in azo dye doped polymer films.

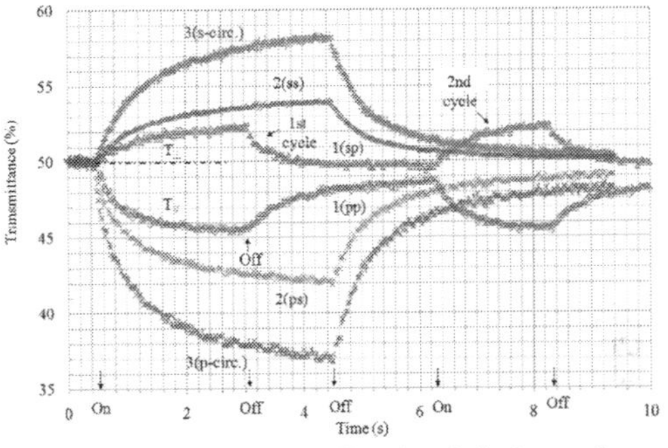

Fig. 6. Circularly compared to linearly polarized pump beam: the photo-induced anisotropy of AZD1 doped PMMA polymer dependence on the polarization state. The first letter after the curve number represents the probe beam's polarization state, and the second letter represents the excitation beam's polarization. Thus, for example, curve 1 corresponds to the vertically (s) rectilinear polarized probe and horizontally (P) polarized excited beams with respect to the incidence plane. For each excited polarization, the excitation starts at t=0,4 s.

On the other hand, it seems that the anisotropy induced by the S-polarized beam is greater than that induced by the P-polarized one. This is not true because the linearly polarized light selectively pump molecules parallel to its oscillating E-field, according to the model of the angular hole burning. Hence, the amount of the induced anisotropy should be the same for S- and P-polarized light, the two cases are equivalent. This should be true whatever the rectilinear polarization direction in a given azo dye doped polymer. In all cases of rectilinear polarization state, the extraordinary axis of the created oblate ellipsoid of refractive indices will be oriented in the oscillating E-field direction of this polarization. Hence, the intensity difference that appears between 1(sp) and 2(ss) curves, for example, is due to the orientation of the oblate index ellipsoid in the 3D space as "seeing" by the circular polarization of the probe beam combined to the effective thickness of the sample.

3.4 Alternately linear and circular polarizations

Fresh azo dye doped film was been pumped alternately by using successively the S-linear and circular polarization states of the same beam under the same experimental conditions. Results are shown in Fig. 7.

Between t=0 s and t=1,5 s, the sample is isotropic. The first cycle in Fig. 7 is assumed to be the same as the first one shown in Fig. 5. The second cycle was induced by using left-circular polarized beam, the third cycle by using again the S-polarized beam and the fourth cycle by right-circular polarization state of the pump. In general, switching off the pump beam leads to a considerable decrease of the resulting anisotropy. Moreover, no significant changes were observed by using left- or right-circular polarized beam (see Fig. 7). Note that the induced anisotropy by using S-polarized beam during the third cycle (in Fig. 7) is slightly higher than that induced during the third and fifth cycles shown in Fig. 5. The third cycle in Fig. 7 was created after illuminating the film by the circular polarized beam while the third cycle in Fig. 5 was after the P-polarized one.

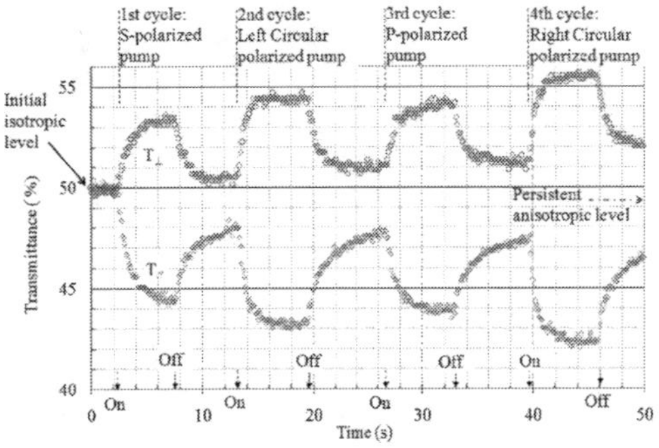

Fig. 7. Four photo-induced/relaxation cycles in a AZD1 doped PMMA film cyclically pumped with On/Off successively S-linearly and circularly polarized beam.

Also, another fresh doped film was irradiated alternately by using successively P-linear and circular polarization states of the same beam. The results are shown in Fig. 8. As one can see again, the circularly polarized pump light not erase residual anisotropy, but induce anisotropy which is important than that created by linearly polarized one.

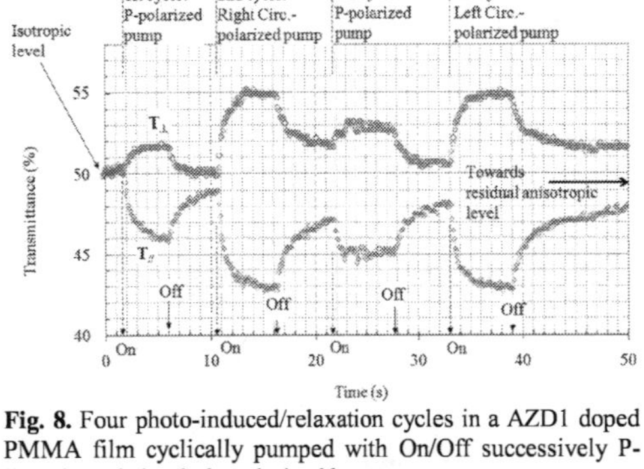

Fig. 8. Four photo-induced/relaxation cycles in a AZD1 doped PMMA film cyclically pumped with On/Off successively P-linearly and circularly polarized beam.

Generally when an azobenzene groups containing polymer film is irradiated with circular polarized light, the anisotropy is erased [8-14]. The use of this idea for almost polymers containing azobenzene molecules continues [1,43-45] and becomes widespread more and more. Our results however show that circular polarized light not only erase anisotropy but induces an anisotropy which is greater than that induced by linearly polarized one (see figures 7 and 8). Natansohn and Rochon [6] are pointed out that if the light is circularly polarized, there will be no preferred orientation within the film plane, but about the same amount of dichroic molecules will probably align in the homeotropic direction. The authors had added nothing to it, but the authors are reported one year before [38] that the progressively decrease of the circular dichroism as a function of number of photo-induced chirality/switching cycles in some azobenzene mesogens in their smectic phase is due to the tendency of the circular polarized light to orient the chromophores perpendicularly to its polarization, i.e., in the direction of propagation of the light. The idea however needed further investigations. Chun-I Chuang et al. [25] are used two different beams from two identical laser diodes at 532 nm to irradiate the Disperse Red 19 (DR19) doped in PMMA polymer. Their results show that the photo-induced anisotropy by superposing two beams with crossed linearly polarization states on a doped film is much higher than that obtained by using only a single beam. Those authors have stated that azo-dye molecules can be further aligned by the extra polarized pump light, thus increase the number of ordered molecules and enhance the induced birefringence. Nevertheless, our results show that during the irradiation with circularly polarized light, the trans molecules tends to orients in a direction perpendicular with the plane of the film which may induce some out-of-plane orientation, whereas in-plane orientation would be minimal since the incoming beam is circularly polarized [20]. Otherwise, the circular polarization state was already used in order to selectively pump molecules parallel to the plane of the sample [18,20] and to accumulate them in the perpendicular direction.

We have modified slightly the experimental setup shown in Fig. 2 to see is there some chirality induced or not. We have inserted a beam splitter (R=T=50%) with high optical quality in the optical path of the pump beam between the $\lambda/4$-plate and the film. This beam splitter (the component symbolized by BS inset Fig. 2) was mounted so that the pump and probe beams are at an incidence of 45° (not the same angle) on the beam splitter. The optical power of the pump beam was adjusted again to 100 mW. The results show that $T_{\perp}\approx T_{//}\approx 50\%$. There is no significant change in the polarization state of the probe beam. Thus, no chirality was induced, according to the literature [42,46]. This observation reveals that the reorientation of the trans species occurs in such a way that the longitudinal molecular axis aligns in parallel with the propagation direction of the pump beam [21,47]. The film was transformed into a positive uniaxial medium whose optical axis is parallel to Oz axis; this is why the film seems as if it is isotropic if we probe it under the same angle of incidence as the pump beam. Anisotropy will not be ersed, only the optical axis will be changed.

Once the circularly polarized exciting light is turned on (see figures 6, 7 and 8), the $T_\perp$ component of the probe increases while its $T_{//}$ decreases. The trans molecules align progressively towards the propagation direction of the pump beam i.e., in the Oz direction; so, in the case of circular polarized light, only the propagation direction is, in principle, perpendicular to the electric vector of the light which rotates around the propagation axis (the Oz axis in Fig. 2) at the angular frequency of the light. Therefore, the refractive index $n_x \approx n_y$ along the Ox and Oy axes respectively decreases to a lower value, while n_z increases to the higher value by comparison to the index of the isotropic sample. The reorientational hole burning [6,40] subsequent to the photo-isomerization push trans isomers to be accumulated parallel to the Oz direction, which is perpendicular to the pumping circular polarization state, as shown in Fig. 2. Thus, doped film does transform from an isotropic to a positive uniaxial medium whose optical axis is in the Oz direction. One obtains an oval index ellipsoid whose extraordinary axis is along the Oz axis and the ordinary axis is in Oxy plane ($n_o = n_x \approx n_y < n_e = n_z$). The experimental setup, shown in Fig. 2, is especially a powerful tool to study the angular redistribution behavior of azo dye doped polymers. Using circular polarized light allows us to pump and probe much better the samples. Taking into account the relatively long live time of cis state which is habitually in the order of few seconds in AZD1 doped PMMA polymer [30-33], the E-field of circular polarized light which rotates at the frequency of the light combined with the polarization state change after each coherence time [48] appears as if it activates the angular hole burning in all transverse directions in the film at each instant: it scans permanently the whole irradiated volume. Thus, relaxing molecules that undergoes the reverse reaction cis to trans will "forced" and not themselves to reorient in the propagation direction of the pump beam perpendicularly to the E-field of circularly polarized light, since only the propagation direction is perpendicular to the rotating E-field vector of the light. Under our experimental conditions, the action of circularly polarized light is to orient the trans molecules along its propagation direction.

Recently, Bin and Oates [49] have proposed a mathematical modeling based on light-induced anisotropic diffusion, in which the trans- (or cis) azo-molecule is considered to be a microscopic dipole, determined by the microstructure (molecular conformation) of the trans (or cis) azo molecule. The results of this models show that in the case of a circularly polarized Gaussian laser beam, the spatial distribution of the trans vector align in the propagation direction of the laser beam. Of course, trans molecules aligned in this direction are usually “invisible” to a probe beam under normal incidence [6,37], as confirmed with our experimental setup (inset Fig. 2). This is also the case of the experimental setup in followings references [35,50] and even many others.

Conclusion

We have studied the behavior of AZD1 doped PMMA polymer by using linearly and circularly pol arized laser beam and also circularly polarized probe He-Ne laser beam. Our results show that circular polarized light not erases anisotropy, but induces anisotropy which is much greater than that induced by linearly polarized light. The use of circular polarized light leading to activate all perpendicular directions of the angular hole burning in the illuminated sport in the polymer film intensifying the angular redistribution and enhances photo-induced anisotropy. Trans molecules align toward the propagation direction of the circular polarized light. The reorientation of azobenzene molecules successively in two orthogonal directions using cyclically linear and circular polarized beams of the light will be useful for applications such All-Optical switching and Optical Data Storage in particularly in the field of nanosceince and nanotechnology.

This work were supported by the - Fonds de Recherche du Québec–Nature et Technologies (FQRNT) - under the grant **FT073270**. The author thanks *Marc-Antoine* Blanchet for help.

References

[1] Y. Zhao, T. Ikeda (Eds.). *Smart Light-Responsive Materials*. John Wiley & Sons, Inc., New Jersey, NJ. (2009)

[2] T. Xu, W. Zhu, Y. Lin, A. Liang. Optik **124**, 305 (2013)

[3] T. D. Mislter. Opt. Phot. News. **March**, 28 (2005).

[4] T. Buffeteau, F. L. Labarthet, M. Pézolet, C. Sourisseau. Macromol. **31**, 7312 (1998)

[5] T. Todorov, L. Nikolova, N. Tomova. Appl. Opt. **23**, 4309 (1984)

[6] A. Natansohn, P. Rochon. Chem. Rev. **102**, 4139 (2002)

[7] P. Rochon, J. Gosselin, A. Natansohn, S. Xie. Appl. Phys. Lett. **60**, 4 (1992)

[8] C. R. Mendonça, U. M. Neves, L. De Boni, A. A. Andrade, D. S. dos Santos Jr., F. J. Pavinatto, S. C. Zilio, L. Misoguti, O. N. Oliveira Jr. Opt. Commun. **273**, 435 (2007)

[9] C. R. Mendonça, L. Misoguti, A. A. Andrade, S. B. Yamaki, V. D. Dias, T. D. Z. Atvars, O. N. Oliveira Jr. Opt. Mater. **30**, 216 (2007)

[10] A. Priimagi, S. Andriy. J. Polym. Sci., Part B: Polym. Phys. **52**, 163 (2014)

[11] A. Goulet-Hanssens, T. C. Corkery, A. Priimagi, C. J. Barrett, J. Mater. Chem. C, **2**, 7505 (2014)

[12] D. H. Choi. Bull. Korean Chem. Soc. **20**, 1010 (1999)

[13] H. Z. Cao, W. Zhang, J. Zhu, X. R. Chen, Z. P. Cheng, J. H. Wu, X. L. Zhu. eXPRESS Polym. Lett. **2**, 589 (2008)

[14] K. G. Yager, C. J. Barrett. J. Photochem. Photobiol. A: Chem. **182**, 250 (2006)

[15] B. Saad, T. V. Galstyan, M.-M. Denariez-Roberge, M. Dumont. Opt. Commun. **151**, 235 (1998). Proc. SPIE, Int. Soc. Opt. Eng. **3418**, 88 (1998). Opt. Lett. **23**, 727 (1998)
[16] X. Ye, M. G. Kuzyk. Opt. Commun. **312**, 210 (2014)
[17] Y. Wu, J.-I. Mamiya, A. Kanazawa, T. Shiono, T. Ikeda, Q. Zhang. Macromol. **32**, 8829 (1999)
[198] Z. Sekkat, M. Dumont. Appl. Phys. B **54**, 486 (1992)
[19] M. Dumont, A. El Osman. Chem. Phys. **245**, 437 (1999)
[20] O. M. Tanchak, C. J. Barrett. Macromol. **38**, 10566 (2005)
[21] S. Sajti, A. Kerekes, M. Barabás, E. Lőrincz, S. Hvilsted, P. S. Ramanujam. Opt. Commun. **194**, 435 (2001)
[22] P. S. Ramanujam, O. B. Jensen, P. Tidemand-Lichtenberg. Opt. Express **21**, 1812 (2013)
[23] B. Saad. Int. J. Dev. Res. **6(11)**, 9946 (2016)
[24] E. Bagherzadeh-Khajeh Marjan, S. Ahmadi-Kandjani, M. S. Zakerhamidi, J.-M. Nunzi. Opt. Mater. **38**, 228 (2014)
[25] C.-I. Chuang, S.-H. Lin, Y.-F. Chao. Opt. Mater. **35**, 366 (2013)
[26] K. G. Yager, C. J. Barrett. Chapter 8, 10-11: in H. S. Nalwa, (Eds.), Polymeric nanostructures and their applications. American Scientific Publisher. (2006)
[27] M. Han, K. Ichimura. Macromol. **34**, 82 (2001)
[28] R. Yin, W. Xu, M. Kondo, C. C.Yen, J. I. Mamiya, T. Ikeda, Y. Yu. J. Mat. Chem. **19**, 3141 (2009)
[29] K. Ichimura, M. Han. Chem. Lett. **3**, 286 (2000)
[30] V. P. Pham, T. V. Galstyan, A. Granger, R. A. Lessard. Jpn. J. Appl. Phys. B **36**, 429 (1997)
[31] V. P. Pham, G. Manivannan, R. A. Lessard, G. Bornengo, R. Po. Appl. Phys. A **60**, 239 (1995)
[32] A. Yavrian , K. Asatryan, T. Galstian, M. Piché. Opt. Commun. **251**, 286 (2005)
[33] A. Yavrian , T.V. Galstian, M. Piché, Opt. Mat. **26**, 261 (2004)
[34] E. Mohajerani, N. H. Nataj. Opt. Mater. **29**, 1408 (2007)
[35] Y. Luo, W. She, S. Wu, F. Zeng, S. Yao. Proc. of SPIE **5642**, 298 (2005)
[36] S. Qi, C. Zhang, X. Yang, K. Chen, L. Zhang, X. Wang, T. Xu, J. Tian, G. Zhang. Optik **115**, 253 (2004)
[37] F. F. Dall'Agnol, F. M. Shimizu, J. A. Giacometti. Chem. Phys. Lett. **608**, 102 (2014)
[38] A. Natansohn, P. Rochon. Can. J. Chem. **79**, 1093 (2001)
[39] F. Zhao, C. Wang, M. Qin, P. Zeng, P. Cai. Opt. Commun. **338**, 461 (2015)
[40] A. Natansohn, P. Rochon, M. Pezolet, P. Audet, D. Brown, S. To. Macromol. **27**, 2580 (1994)
[41] A. Natansohn, S. Xie, P. Rochon. Macromol. **25**, 5531 (1992)
[42] P.-A. Blanche, Ph. C. Lemaire, M. Dumont, M. Fischer. Opt. Lett. **24**, 1349 (1999)
[43] J. Vapaavuori. Doctoral Dissertation. *Design of efficient photoresponsive azobenzene materials through supramolecular functionalization.* Aalto University. (2013)
[44] A. Priimägi. *Doctoral Dissertation. Polymer-azobenzene complexes: from supramolecuar concepts to efficient photoresponsive polymers.* Helsinki University of Technology. (2009)
[45] F. Filippo. Ph.D. Thesis. *Photoinduced deformation in thin films containing azo-derivatives.* Ecole Polytechnique, ParisTech. (2009)
[46] Y. Kalachyova, O. Lyutakov, I. Goncharova, V. Svorcik. Opt. Mat. Express **5**, 2761 (2015)
[47] A. Yavrian, T. V. Galstian, M. Piché. Opt. Eng. **41**, 852 (2002)
[48] E. Hecht, Optics. 4th ed., Addison Wesley, New York, NY. 2002
[49] J. Bin, W. S. Oates. Sci. Rep. **5**, 1 (2015)
[50] T. Xu, G. Chen, C. Zhang, Z. Hao, X. Xu, J. Tian. Opt. Mat. **30**, 1349 (2008)

*****

AUTHOR INDEX

EDP Sciences – Web of Conferences
17, Avenue du Hoggar
Parc d'Activité de Courtabouf
BP 112
F-91944 Les Ulis Cedex A
France

ISBN 978-1-5108-3834-5

CURRAN ASSOCIATES INC.
proceedings.com

Ninth International Conference on Cloud Computing, GRIDs, and Virtualization (CLOUD COMPUTING 2018)

Held at ComputationWorld 2018

Barcelona, Spain
18 - 22 February 2018

IARIA